SOAS Studies in South Asia

The Unhappy Consciousness

BANKIMCHANDRA CHATTOPADHYAY AND THE FORMATION OF NATIONALIST DISCOURSE IN INDIA

SUDIPTA KAVIRAJ

DELHI
OXFORD UNIVERSITY PRESS
BOMBAY CALCUTTA MADRAS
1995

Oxford University Press, Walton Street, Oxford OX2 6DP
Oxford New York
Athens Auckland Bangkok Bombay
Calcutta Cape Town Dar es Salaam Delhi
Florence Hong Kong Istanbul Karachi
Kuala Lumpur Madras Madrid Melbourne
Mexico City Nairobi Paris Singapore
Taipei Tokyo Toronto
and associates in
Berlin Ibadan

ISBN 0 19 563294 X

Typeset by Rastrixi, New Delhi 110070
Printed in India at Rekha Printers Pvt. Ltd., New Delhi 110020
and published by Neil O'Brien, Oxford University Press
YMCA Library Building, Jai Singh Road, New Delhi 110001

Contents

Preface

It is always difficult to make a proper accounting of intellectual debts. In writing this book my greatest debt is to Partha Chatterjee for showing me his work on nationalism in its very early stages. It showed me how interestingly nationalist texts could be re-read. He also convinced me that what I had to say about Bankimchandra should be tried out in a separate work rather than diffused in endless conversation. Tapan Raychaudhuri similarly encouraged me to think out an argument about Bankim's treatment of Krishna. I also remember with gratitude Bimal Krishna Matilal's unfailing encouragement at different stages of this project.

As the book grew out of conversations and lectures, it was shown at various stages in different versions of untidiness to people who were generous with their time and gave me their critical reactions. David Arnold, Jasodhara Bagchi, Gyanendra Pandey, Hitesh Sanyal, and Asok Sen read parts of the manuscript and made helpful comments. I greatly benefited from discussions on various points with Veena Das, Ashis Nandy, Ranajit Guha and Namwar Singh. I owe special thanks to Ravinder Kumar for his encouragement and the provision of the excellent facilities of the Nehru Memorial Museum and Library during the writing, and for taking the drafts into their Occasional Papers series for limited circulation and discussion. The Centre for Studies in Social Sciences, Calcutta, invited me to be a visiting professor for a summer, and provided the ideal intellectual setting for trying out some of these ideas in a preliminary form in the intense and enjoyable discussions which followed the lectures. Several parts of this work were presented in seminars at the University of Delhi, Jadavpur University, Panjab University, Chandigarh, University of Heidelberg, and St Antony's College, Oxford. I have learned a great deal in political philosophy in many years of association with Rajeev Bhargava and Sanjay Palshikar, and although I could not share the texts of Bankim with them directly, I am sure my way of seeing the texts owes a great deal to the ideas that they have generously shared with me.

1

A Taste for Transgression

Liminality in the Novels of Bankimchandra Chattopadhyay

In his work an artist creates a world. Two distinct types of questions can be asked about this world of his. First, what kind of a world is this—what is its structure, its limits of possibility, the inner logic of its working? Second, how did this world that the artist created relate to the world in which he lived?[1] In thinking about literature it is difficult to keep these two types of questions clearly apart, but they are analytically distinct. The second set of questions represent more the historians' concerns; the first is textual, that is, the answer to them must be found within constructs that are internal to texts.[2] This essay seeks to understand the structure of the world in Bankimchandra's fiction.

It can be argued that to speak of 'the world' in both cases is misleading; because the two are not worlds in the same sense. Indeed, realists may assert that it is only the second which really qualifies to be a world, by virtue of something like the doctrine that the world is all that is the case. There could be disputes about whether the world of the artist is part of all that is the case or not. That in turn will raise questions about the ontology of the work of art and the language we use to describe them.[3] Without going into that class of questions, we can still see that even if both are 'worlds', even if they are both 'real', they could neither be real nor worlds in quite the same sense. What I wish to stress is that it is seriously wrong to deny the createdness of this world. There is something odd about the extreme realist position invoked often by historians rummaging in literary texts for historical material which holds that art can only reflect or play variations on reality. Indeed, the idea of 'playing a variation on reality' is a paradox for the realistic theory of art to solve. For the idea that one can 'play with' reality implies that one need not take it as given, and suggests the possibility of an alternative relation to it. What we try to see in the artist's art is not what he sees, but the way he sees it. His way of seeing the world cannot be an object of *his* sight; but for others that is the proper thing to see, for otherwise we lose the distinction between art and the world.[4]

As a serious artist, Bankimchandra created an artistic world of great internal variety. He was extremely sensitive to possibilities inside language, and controlled its resources with an admirably deliberate precision. He could create inflections at the touch of a word—a skill which may have come down to him from the classical Sanskrit tradition he admired, in which apart from a fictive imagination, a highly refined linguistic sensibility was considered a prerequisite for literary creation. Bankim's world has a sunny and a dark side. The lighted side of his world is more easily seen, because he was one of the most accomplished writers of humour in the language.[5] It is not proper, however, to see his humour as merely an inverted form of a philosophical melancholy. It seems to me that the unifying vision, the underlying experience which runs through and structures his fictional art reveals a dark vision of the world. By attending to this aspect of his art we might be able to see a structure in his world-view which may help us understand better his mind and its engagement with history. Some themes and problems occur in his novels too frequently to be dismissed as merely incidental attributes. Indeed, I think it can be shown that one of these central problems is constantly refined, given a central and clearer structure as his art evolves, moving from the margins of his artistic curiosity to the narrative centre. This is a concern with the liminal.[6] Bankimchandra must have been long intrigued by this problem of liminality; and must have felt that he was in the presence of an ontological question.

Bankim's central philosophical concern could be described in two ways. Stories of his novels often turn around a conflict between two inevitabilities, two things that are equally necessary truths of human life. A social world required definitions, a kind of basic social map which defined permissions and prohibitions. At the same time, there are the elemental drives of human nature which these social constructs are meant to discipline into reasonably safe forms, but hardly can. The social and moral worlds in which men actually live are made up of these two dissimilar and contradictory elements—the desires which make men and the controls which make society. Much of Bankim's fictional movement arises from this central conflict—between the inevitability of moral orders and the inevitability of their transgression.[7]

This could be seen in another way. Bankim's novels almost always centre on a play around the liminal, usually a liminality of morals, constructing fictional situations which pose large, often unanswerable, questions to constituted moral theory. But this concern with moral liminality is not the shallow kind of moralizing one often found in Victorian novelists, who painted a morally placid social universe to frame a drama of muted desires, where transgressions are mere misunderstandings. At the end of the story the symmetry of the moral order is restored. Bengali fiction by and large accepted the narrative norm of this Victorian type.

Bankim's concern with morality is of a very different kind. Morality is of wonder to him because it seems to touch the ontological, to reveal a romantic and fearsome darkness at the centre of the world.

When the first shape of this fictional world was being etched in, at its moment of creation, in the first scene of his first novel, his first character saw the world in a dark light. It is not his seeing that made the world dark; that was the way the world was. What Jagatsinha saw in the darkening evening in the opening scene of *Durgeśnandinī* was, I suspect, more than an external landscape; it was a glimpse of the nature of the world. It did not represent something contingent, it expressed a necessity.

> The night fell before he could cross the immense distance of the field. The sky grew dark with black clouds. The clouds of night made the horizon so dark it was difficult to ride on. The traveller made his way in the erratic light of the occasional lightning. Soon a violent summer storm broke out with great violence, and a thick rain began to fall. The horseman lost all sense of direction and did not know which way to proceed. . . .

As we know, he eventually blundered in this rainy night into an encounter with his destiny. That is how human destiny is met.

The way he meets his destiny—the face of Tilottamā—is symbolic. He meets her in an abandoned temple, in a place of ambiguous holiness, in secret, in the flickering uncertain light of a lamp which soon went out in a gust of wind. What a way to meet one's destiny, to have the first glimpse of the face which would change one's life! Or is that the way it is met, always? Anyway, it was a fatal landscape. Its gloom and foreboding could have come from one of the romantic paintings of the unequal meeting of man and the forces of nature. Like them, it seems to bear a dual significance. It was not merely an event occurring in an individual story, not the destiny of *this* individual man, nor the darkness of *this* rainy night. There was another night which stood beyond this night, another darkness beyond its darkness. As in the paintings of the Romantics, the contingent, the event depicted, was significant precisely because it expressed something larger than itself, it revealed the nature of the world. It was a metaphysical inscription; and the storm of that night was the storm of existence. It was also a landscape of the creator's mind. This is how he thought the universe was—tense, disorderly, nocturnal, swept by incomprehensible disorders. It was not accidental that facing that disorder was a man, a representative in Hindu theory, of the principle of order. This was an intimation of the condition of humanity.

The general question of liminality seems to have fascinated Bankim as a novelist, though the form of this liminality, its social context, varied a great deal. In *Durgeśnandinī* the moral-personal aspect of the plot is remarkably daring. Jagatsinha, the protagonist, is seen at the intersection of two evolving relationships: his own love for Tilottamā and Āyeśā's growing love for him.[8] Sometimes it is suggested that the boldness of the plot is derivative; for the similarity of its narrative structure with Scott's *Ivanhoe* is too plain. I find an irony in the title of Bankim's novel: *both* women can be called equally appropriately the daughter of the master of the castle.[9] Āyeśā overshadows Tilottamā and Jagatsinha. It is her passion for Jagatsinha which holds the story together, makes it move, to use Todorov's phrase, from the first adjectival equilibrium to the second.[10] The play about liminality is her play.

The concern for liminality seems to me quite central to Bankim's plot: what was interesting to Bankim the novelist is the fact that the ways of love are inscrutable, and the possibility that it *can* arise unbidden between people who are separated by an uncrossable social distance. The limits which separate Āyeśā from Jagatsinha are not barriers placed there by the war; these are deeper prohibitions that separate two religious communities. Yet that does not stop the possibility of romantic passion from really existing. The contradiction is between two aspects of the relation—what is humanly possible is socially unfeasible.[11]

The structure of *Durgeśnandinī* demonstrates that its narrative is a play on liminality, a particularly dangerous one for a first novel, so much so that even Bankim did not play with it ever after, which even he in his maturity, perhaps, considered too dangerous. The way Bankim's thought moves after *Durgeśnandinī* can be seen as both a *radicalization* and a *deradicalization*. Seen from one point of view, his fiction construction gets less radical, in the sense that the limit between two religions raises the spectre of a great transgressive act, much greater, if it had been crossed, than liaisons with widows or otherwise desirable but unattainable women within the confines of Hindu society. In that sense, Bankim retreats from this kind of a radical theme. In a different, more formal sense however, he radicalizes his problem, because what appears as a mere possibility, a mere hint of a storm, only the *idea* of a transgression, becomes in later novels a series of *real* transgressive acts. Āyeśā does not enter into an actual transgression; Bankim's subsequent heroines all do: Kunda, Rohinī, Srī, Indirā all enter into real transgressive acts. What causes this play around liminality is the conflict between desire and denial, desire seen as man's (or mostly woman's) elemental inclinations, and denial as the system of prohibitions constructed by society to bind and channel them, and render them safe. In the world of his creation, which he slowly constructs with thirteen other novels after *Durgeśnandinī*, this struggle between the brittle creations of society and the waves of human desire has the central place. However, its form changes. Characters who enact this drama vary. The play is constructed differently in each case; but there is an underlying unity of philosophic concern about this question, this dark possibility of transgression.

I have earlier called this a philosophic problem stemming from morality. It is important to see exactly what this means. It is not a moral problem in the narrow sense. A moral judgement often judges an act or a life against rules set down by a particular moral system, and evaluates them ethically as right or wrong. But the kind of judgement this type of story-telling invites is very different. What we come to reflect upon and judge is less the moral quality of an act than the nature of morality itself, and the rightness and wrongness of having this kind of a system of right or wrong. The feeling left in us at the end of the story is not one of moral horror at the presumption of Āyeśā's transgressive love for Jagatsinha, but a stealthy fundamental sympathy. To use another philosophic term, we get a tragic taste of finitude. It is not an empirical question of judging an act against a set of moral rules; but reflecting upon the necessity of having to judge, of having moral codes which prohibit inclinations, impose an impoverishing format on human lives, destroying happiness by interdicts.[12] We come not to say how wrong Āyeśā's emotions are, but how wrong is a world in which such feelings have to be seen as wrong. It creates—to use Rene Girard's argument—not a sense of wrongness of the act, but of wrongness of the world—the basic emotion of the tragic. In one sense, it explores the behaviour of a 'problematic individual', but by the skill of art it accomplishes a subtle transformation of this enquiry, and turns it into a sense of the problematicity of the world itself.[13]

Liminality of a related but somewhat different sort arises for the first time in Bankim's third novel, *Mṛṇālinī*,[14] a story without much narrative complexity. Two stories are woven together simply by the accident of time and space. The protagonists of both stories happen to live in Nabadwīp when the events of the narrative unfold. An outer structure is formed by the title story of Mṛṇālinī and Hemcandra, the deposed crown prince of Magadh, seeking revenge against the Muslim despoilers of his father's kingdom. His character does not display any remarkable quality except a muscular unintelligence which causes Mṛṇālinī much avoidable suffering.[15] Within the story of Mṛṇālinī and Hemcandra, there is an inset of a remarkable second narrative which reveals all the basic elements of Bankim's metaphysical search for the nature of the liminal.

Gaud, an independent kingdom of Bengal, is ruled formally by its feeble old king, Lakṣmansen. Actually, the man who rules in name of the old king is Paśupati, an extraordinarily able Brahmin, a schemer, above ordinary moral scruples, constantly plotting violent deeds for the sake of power. At the centre of his personality is an intense and interesting frustration. He loves the daughter of a local priest, of vague antecedents, Manoramā—a woman of surpassing beauty. (A constant feature of Bankim's fiction: I think it is not just a matter of literary conventions which did not allow unspectacular women to be objects of art; they had to be natural masterpieces to find a place in artistic ones. But there may be a deeper philosophic point: they have to symbolize the power of their sex, their fatal attraction; for they always represent the invitation of the liminal; their beauty is a mark of their being expressions of a principle, something larger and more inscrutable than themselves.) Nothing is known of Manoramā's life except that she had lost her husband and was a young widow, the symbol in Hindu society of both attraction and interdiction. Bankim invests her figure with all the mysteriousness that the Hindu religious and cultural traditions characterize a woman with—a contradictory personality fluctuating between the serious and the playful, a clear intellect and childish innocence and, above all, a masterful capacity to use precisely this duality of her personality to keep Paśupati at bay.[16] In the end, Paśupati dies, unable to control his destiny though he was perhaps the most competent of rational players of his time, showing to us one face of fate.[17] Paśupati experiences a consuming passion for Manoramā, which does not bring her to him. When he dies, a victim of his own diplomacy, she is still distant, unattained. At the end of the novel, it is revealed that Manoramā was not a widow; she was in fact, Paśupati's wedded wife; but he was kept in the dark about this by a false interpretation of an astrologer's forecast.[18]

Here we find a man and a woman bound together by two distinct relations simultaneously, and the relation between the two relations is contradictory. The first is the relation of marriage, the socially sanctified form of desire. The second is a socially unsanctified form of passion, something dark and nocturnal, something that threatens the mapping and the whole architecture of the social world. Obviously, these two relations symbolizing two forms of desire, are contradictory in the social world that Bankim took as given, the map of moral order provided by the Hindu tradition. Every structure of its kind is defined by its basic antinomies, the central definitions which must be held in place if the entire structure of secondary definitions is not to collapse in confusion. Certainly, within Hinduism the exclusion between wife and mistress is an antinomy of this kind. Within this structure, they are exclusive of each other, in the sense that if one exists, the other cannot. The same woman cannot be both at the same time to the same man.[19]

Seen from this angle, Paśupati is actually Manoramā's husband, but he is also, at least potentially, the dangerous transgressor. He should have had Manoramā as his wife, but is reduced by circumstances to desire her as a mistress. In a sense, therefore, in this curious story we find an overlay, a confusion of the two antithetic definitions—a woman who is a wife is desired as a mistress, though, in retrospect, in comparison to *Indirā*, a later novel, this particular narrative-philosophic construction is not radical enough. The only flaw in this extraordinary construction is that it has to use the condition of Paśupati's ignorance, and that his desire for Manoramā remains unconsummated. Paśupati desired her, but it remained a mere desire; he did not fall into actual 'sin'. We shall see that this flaw would be rectified in *Indirā* by a more radical construction of the narrative.

It seems to me that what is implicit in this extraordinary plot is an artistic enquiry into the ontology of social relations. Is not the artist asking through the story what social relations are, and what constitutes moral judgement about them? But what kind of thing is a relation of this social-moral kind? After all, the sexual relation between two individuals is a fairly simple affair. Its mysteries are not philosophical. It is free of the problematic and elusive quality of the social descriptions that could be given of it. Moral and immoral are social descriptions. They presuppose a society which chooses one universe of moral discourse among many possible ones. Moral descriptions, being wife or being mistress, because these, after all, are not mere states of affairs, are not such simple things. What is really added on to the sheer behavioural relation between two individuals for it to turn into something morally acceptable or condemnable?

Some would perhaps object that this questioning of liminality was not part of the novelist's intentions. Intention, however, is an odd and surprisingly complex matter, particularly when applied to artistic texts. Surely it is odd to suggest that such a momentous questioning of the limits of the moral order he accepted was an entirely non-deliberate affair, that he did not see in this radical sense what he was doing. An accident that happens in seven novels is an extraordinary accident indeed!

What is sometimes said of painters and artists is also true of novelists, and of Bankim. Artists often seek a perfect realization of a problem or an image. Leonardo, Freud said, sought the perfect rendering of that mysterious beginning of a smile—through the remarkable consistency of his painted women.[20] Bankim seems to have had a similar obsessiveness with the problem of duality of relationships, i.e., the same empirical relation being eligible for two contradictory descriptions, until he reached the clearest and the most extreme form in the otherwise unspectacular novel, *Indirā*.

By common consent, *Indirā* is not one of Bankim's best novels, whether judged in purely aesthetic terms or by the normal criteria of excellence of the novel form. The story which holds these incidents together is highly contrived, a serious flaw in construction given the ideals of realist art that Bankim consistently pursued. But his is a prenaturalistic realism. Linearity of narrative is the norm here; he indulges in an old-fashioned story-telling, quite distinct from the perspectival experimentalism of some of his own later novels like *Rajanī*.[21] The basic trouble with the story is its unreality. One can take it seriously only through a willing suspension of disbelief on crucial points. Even if the robbery is credible, though the daughter of a wealthy family would perhaps go out somewhat better protected in those unsettled times, Indirā's subsequent run of uninterrupted good fortune is hardly so. Indirā is lucky in meeting a Brahmin who apparently exists for the explicit purpose of helping girls in distress. She is lucky in getting an employer like Suhāsinī. She is incredibly lucky in the chance of her husband's susceptibility to physical beauty and her possession of the exact kind that flatters him, and particularly the fact that this potential was not already exploited by any preceding cook in his life.

Yet it is difficult somehow to say that *Indirā* is entirely unrealistic. One finds pleasure in Bankim's characteristic observation of the small things of domestic life. The realism of a story is, after all, a rather complex affair. It is not just a matter of the plot, the sequence of events which constitute the movement of the story from one point of equilibrium to the next. If we use Todorov's distinction, the story is unrealistic in its 'verb' aspect, but entirely credible in its 'adjectives'. All those circumstances which move the story from one equilibrium to another are unconvincing, but the states of equilibrium which it depicts, when the story does not move, are characterized by an undeniable social realism. It is an entirely unlikely chain of events in an entirely likely world.

One reality it depicts with particular conviction—the world of women in the ordinary *bhadralok* household, and how different the women's sense of the world is from that of the men's which is ordinarily taken to be the self-evidently 'right', normal view of these things. It shows consequently, the obliqueness of the feminine vision. In this sense, the skilful use of language in the dialogues plays a very significant part; and not accidentally, dialogues constitute a large part of the novel. Justifiably, *Indirā* will figure prominently in any discussion on the development of the Bengali literary language. Bankim's fictional work is marked by an astonishing openness and experimentalism in language use. Quite deliberately, he gives us a reprieve from the conventional one-dimensionality of fictional language, the monotony of a novel all written in one language—either the formal or the colloquial.[22] His language is always full of surprises. It changes in its key, its tone, its refinement and coarseness, its symbolism. Indeed, it is quite distinct from the language of other contemporary writers in its ability to use coarser linguistic forms selectively, something that Tagore, for instance, never does. Language inside his fiction shifts constantly between levels, changing with great sensitivity to contexts. Thus the play of language adds to the constant play of the plot, between formal and informal language, between high and low, the language of the ceremonial and the spontaneous, of men and women, of the home and the office, languages of description and of dialogue, the author's language and the characters' language, the character's language varying according to what kind of character it is. Curiously, there are two ways in which language could be said to have been used well in fiction: one is a subtle and latent use where the aptness of linguistic expression makes us nearly forget the presence of language, where language makes its presence transparent. The other style presents language with an unforgettable tangibility where the 'rustle of language', its sounds, its play, its surprises, never allow us to forget that what is happening is happening inside language where the semantic is never able to eclipse the aural. Bankim's usage is of course unmistakably of the second kind. Through language Bankim is able to realize in the novel the kind of change of scene and alteration of perspective one usually finds in a drama, in the constant first-personness of the dialogue. In a purely formal sense, Bankim's novels have a mix of drama. His fiction does not have the inevitable narrative length of novels but has the compactness arising out of a condensation of action and movement more akin to drama. This technique of language use however was not new. In this he was following an established tradition of classical Sanskrit plays. In *Śakuntalam*, for instance, there is a careful matching of character and the weight of language used, moving from the ornate formal language of hermits and the king to the vernacular *prākṛt* spoken by women. Evidently, this performs a dual function: it makes conversation realistic, but it also works as a kind of linguistic indexing, alerting the audience to the symbols of status.

Before Bankim, in Bengali fiction two types of language use were common, both, in the sense I am using the term, one-dimensional. A more conventional style could be found in Vidyāsagār who used a heavy, formal language for both narrative and dialogue, the language of the narrator and of the characters, an internal and external language.[23] There was however a dissenting tradition in the use of colloquial Bengali in *Alāler Gharer Dulāl* and *Hutom Pencār Nakśā*.[24] Both types however stuck to the unity of the internal and external language and are quite different from Bankim's style.[25] He uses a heavy, formal, but extremely resourceful language for the narrator—which could reach into the riches of Sanskrit grammar; and a colloquial idiom for his speakers, varying this context sensitively. By implication, this creates a distinction between the spontaneity of life and the conventionality of narrative.

Context-sensitivity of language used in fiction is to be seen in two ways. It is sensitive to both situations and actors. Language used by men and women is idiomatically different. Apart from the fictive pretence of *Indirā* being an autobiographical novel, and therefore the language of the narrator being indistinguishable from the language of the central character, the novel is linguistically most remarkable, for it takes the culturally neglected colloquial Bengali of the kitchen[26] and places it right at the centre stage of high literature. In that sense, it certainly constitutes a landmark in Bengali fiction. This leads to an internal transformation of the genre Bankim had founded. It is a dramatically undramatic novel, a startling departure from the Gothic plots of *Durgeśnandinī* and *Mṛṇālinī*, and equally distant, though in a different way, from the tales of nemesis within his social novels. *Indirā* is an undramatic novel except in its unconvincing first section and the even less convincing last act; perhaps its drama is of a different kind. Indeed while the others are novels of action, this is a novel which explores relationships, in which actions are negligible and external. What happens goes on in the kitchen or its close vicinity. At any rate, it rarely moves out of the *andarmahal,* the women's sanctuary within the Bengali household, except for some forays to mock at that lugubrious symbol of the formal, westernized outside world, the perennial object of Bankim's ridicule—colonial courts.

Some interesting allegories could be found on this level. There is at play a caricatured version of the philosophic distinction between male and female principles, a serious enough idea in original Hindu philosophy but adapted by the imitative Victorianism of the *babu* into one between an all-important outside world and an unimportant home. The babu version, the rationality of the male and sentimentalism of the female, is of course a caricature of the original traditional idea; and there is evident justice in its being caricatured in turn. It shows up the world-consuming pretensions and delusions of the babu, the ineffectuality of the subordination he wishes to impose on others who live around him. Men represent, in this very masculine of myths, the outside world of weighty and intricate decisions, women the uncomplicated sentimentality of the household. It is men who hold the world in order, and stop it from dissolving into a confusion of indulgence and tears, a stereotype that middle-class women commonly internalized and enacted. Bankim wished to ridicule this simple dichotomy of the babu by asserting a Shaw-like thesis of women being the prime, instinctual, if indirect, makers of decisions. This may, at a deeper level, be related to a point about the play of subalternities of various kinds. Bankim gestures often in his novels, to the abjectness of the babu, his utter inconsequentiality in the office world of British administration, a world which he can give out to be his only by dissembling. For this babu, this sovereign rationalist master within his household, is a Mucirām Guḍ in his office. He can hide his social subalternity only on condition of the woman's domestic subalternity, on condition she does not know anything about the world outside. Still, often she knows; she sees through his pretence of rationality, control and power.[27] But she is a creature of the repressive Hindu middle-class household, and is adept at the political use of pretence against male power. Her subtlety creates a situation in which the babu does not understand that she understands. This leads to an ironic inversion of relations. Eventually, it is not the voluble westernized rationalist male who represents the principle of autonomy; it is she who, inside her household fortress, remains hidden and culturally unconquered.

Language is thus intimately related to the social contrasts in the complex world that stretches between the kitchen and the court. In the world of his novels, this dichotomy stands for many things; several other dichotomies are superscribed on this basic scheme. Men are more formal, ceremonious, often to the point of being ridiculous; women are spontaneous, without worldly pretensions, a contrast reflected with particular clarity in the many unequal contests in *Kamalākānter Jobāndandī* between the *goālinī* Prasanna and the ponderous lawyer, to the detriment of the latter.[28] Men are unpractical, though occasionally idealistic; women are entirely practical, with a winning combination of a practical generosity and a perfectly understandable cynicism about the effectiveness of high ideals in a corrupt and insensitive world. Women here are almost everything they are in Shaw's drama—spontaneous, clever, witty, adaptable, flexible, with a winning changefulness. All this is reflected in the deliberate counterpointing of the two languages men and women speak, especially in the unmasking done by devastating redescriptions of a formal act in ordinary language.[29] Women's language has fluidity, caprice, effervescence, spontaneousness, a native wit that sets it apart from the more formal, neutral language of the narrator and the male language—ponderous, formal, deceiving. Women too deceive, of course; but they deceive beautifully and well, and in insignificant matters. *Indirā* is, in a sense, a celebration of this language more than anything else. It gives off a sense of enjoyment of its freedom, its immediacy, sensuousness, its nearness to life. As it is a distinct language of a distinct world, through its use people are given a glimpse into this world, as if the curtain has flown up and a furtive glance stolen into the mysteries of the inner courtyard, of the sorrows and happinesses of this world, so completely stifled by the officious exterior, so near to the male world, yet so immensely distant. Bankim penetrates with far surer intuition into the woman's everyday life than does Tagore in his far more formal study of the tensions of a woman's world in *Ghare Bāire*. The difference between Bimalā and Indirā is interesting. Bimalā is a liberated woman in many respects, but in a liberation defined by her husband as a representative of the current upper middle-class culture and worked by him. She grows into a second subjectivity of her husband, rather than a subjectivity of her own. This certainly represents an achievement of autonomy, but this autonomy is defined in someone else's terms. Indirā is also the picture of an unconquered woman, but in a very different way. Indirā's autonomy shows the irreducibility of the woman to men's standards; like some of Bankim's other ladies she refuses to fall into a notion of cultivation defined by others. But these characteristics of the novel, though interesting, are not its central interest.

As I wish eventually to distinguish between a surface and a deep structure in the story, and since the deep can be reached only through an initial resolution of its surface properties, let us look first at what the narrative offers on the surface. Indirā, the daughter of wealthy parents, is going to join her husband. It was customary for young girls to be formally wedded; but often they lived in the parental household, and moved to their husband's family in late adolescence. On the way her small entourage is set upon by a gang of exceptionally law-abiding robbers. Their sense of the limits to looting practices introduced by the English rule of law is admirably precise. She is robbed of everything including her expensive sari. An inexperienced and youthful robber suggests that she is too pretty to be left behind; but a senior colleague advises professionalism, and demands that they do not mix business with pleasure. After all, they were now part of the efficiently administered Company regime. A rationalist sense of social differentiation and strict rules of profession had spread to even the more aggressive trades. Common robbers in Bengal's forests were not immune from the spread of European rationalism. Thus let off, she hides in bushes until dark. At night she goes to a Brahmin family who take pity on her, and on the supposition that she must come from a distinguished family, send her off to Calcutta to find suitably undemeaning housework. In Calcutta she is given the non-menial work of a cook in a lawyer's household. The lawyer's wife guesses that she must have some dark and tragic secret, worms it out of her and willingly believes in her improbable story. She resolves to reunite Indirā with her husband. Happily, in this best of all possible worlds, Indirā's husband is a client of this very lawyer; inscrutably kind are the ways of providence if one is born into the right kind of family. Indirā's husband is called to Calcutta for an ostensible consultation on legal matters. He is served food by the alleged cook, who is careful to let him have enough of a glimpse of her to be absolutely smitten by her beauty. He is so struck that he proposes a rather sordid arrangement for such a noble sentiment—she should live out of wedlock in his house in Calcutta, a fairly common expedient for affluent Bengalis of the time. Indirā agrees to do so, and during this period of cohabitation lets her husband gather through her conversation that she is either his lost wife or someone endowed with supernatural powers. At the end she is united with her husband through irrelevantly unconvincing means, her 'double' husband too fond of her to scruple excessively about the vagueness of her earlier career.

Under its thin and unconvincing narrative plot the story poses a *philosophic* problem. Since we can detect a genealogy of this question in Bankim's earlier work, such a hypothesis becomes more credible. Even to the non-philosophic reader the story must have underlying compulsion because of the masterly way it handles the question of identity. Indirā is originally wife to her husband Upendra. By the robbery, she is taken out of the map of socially established moral relations. She, in other words, loses her claim on Upendra, because that claim is dependent on some conditions she is not able to fulfil. (What is interesting here is the similarity of her case to the mythological figure of Sītā. She had not transgressed; but the point is not whether she has in reality transgressed or sinned, but the impossibility of proof to a sceptic that she has not.)[30] Her only way of getting him back to her, which is the same as finding her way back into the social map, is a more elemental way, to use her sexual attraction, which she uses, happily, with fatal effect.

In all this her identity is under radical questioning; and, I shall suggest, what gets interrogated is not the identity of an individual, but the nature of identity.[31] What makes one socially what one is? Indirā's identity is shown to be full of contradictions. She is a wife who ceases to be a wife, which raises the question, is the relationship of being a wife something which can cease to be? In the middle of the story, Indirā still regards herself as Upendra's wife. But the title of a wife is a socially conferred one, and her behaviour shows that she accepts that she does not have that kind of claim on her husband. Clearly, the relation she comes to have with Upendra is a sinful relation, of being a mistress. Upendra notwithstanding his fondness treats her accordingly. The philosophical point of the story lies in the duality of their relationship. From Indirā's side the relation is still one of marriage, only accidentally suspended, a status she is rightfully trying to recapture. But it is not really that simple. For there is also a sense in which she implicates her husband in a relationship that is really sinful, because it is extramarital. The relation between these two relations—of being wife and being mistress and their alleged incompatibility—is the most interesting question of all.

On Todorov's schema, in this narrative, there are three distinct points of equilibrium. In the first, before Indirā sets out on her fatal journey, she is of course Upendra's wife. After the incident with the dacoits, this identity as wife lapses. A marriage relation, this shows, far from having objective attributes (and despite being made in heaven) is a social construct—in fact, of a particularly vulnerable texture. This is a problem that is raised and interrogated constantly in the classical Indian literary tradition from the *Rāmāyana* and *Abhijnāna Śakuntalam*. At the end Indirā wins her status back but entirely by her human device, and by wholly questionable means, not by the omnipotence of the sacred tie.[32] Marriage ties are doubly vulnerable: if they are not recognized, they simply cease to exist; secondly, the marriage tie can be destroyed with astonishing ease by an elemental desire. Indirā cleverly uses a force which normally breaks marriages to reconstruct hers. Finally, of course she becomes that wife again; or does she? Whether she does or not is a question which is not raised because of narrative closure. She gets back her due, strictly speaking, through sin. If knowledge is a constitutive aspect of moral behaviour, for her husband the affair with this nameless cook is a gross transgression. Upendra is merely another of Bankim's gallery of irresolute men who succumb to the seduction of the flesh, another victim of the irresistible sexual attraction which society sanctions with a feeble seal of approval. On Indirā's side too, the behaviour is at best morally ambiguous. Since she knows that her husband does not know that they are married, it is hardly unimpeachable conduct on her part. She can however claim a desperate extenuation. She is driven to it by a kind of Hobbesian argument—that there is a grading of moral occasions, that even something which is not ordinarily justifiable is right for the sake of self-preservation.[33] She is driven to do this to win her way back into her own life. For she has become a kind of refugee from her own self, seeking herself from the outside.

Still, it is not easy to define her relationship with the man once the two relations merge into one. It cannot be the formal relation it started with. The bond between them has meanwhile become stronger because it is more elemental. It cannot be the same, secondly, because it has also become tainted. This is the first level of irony of the narrative. The socially sacred bond of marriage is brittle; the attraction of sexuality, a taboo, has an irresistible elemental force; and Indirā invokes the low in aid of the high. This also reveals the underlying question about the ontology of social relations—how do the two relations relate to each other? What kind of a thing is a social relation really?

The complete inextricability of the relations shows the second paradox: is it possible to have a situation in which a woman in order to be a wife must become a mistress? What I wish to convey through the idea of indescribability is this: if we say Indirā is Upendra's wife, this will be false, because this will leave an important aspect of their relation undescribed; but this is equally true of the alternative possible relation of being a mistress. A proper description of what she is to Upendra could be given if we had a term which could cover both wifehood and mistresshood. But this is ruled out by the map of possible moral relations in Hinduism. They are contradictions; consequently, there could not be a single description which could cover both. If such a term could exist, the meanings of the two original terms would have collapsed into confusion. Thus the construction of a story which, from the point of view of realism, appears improbable and aesthetically forced, in which the only interesting thing was the woman's view of the world (of course there are problems here too associated with male construction of feminine subjectivity) comes to reveal an interesting philosophical and moral problem. There can be situations which defy the mapping of social or familial space. In such cases, the usual, practised judgements are reduced to silence.

A comparison with Manoramā's story (in both senses, i.e. what happens to her, and the way it is told) is interesting. Obviously, in some ways Indirā is an intensification, a radicalization of the paradox we found in the earlier novel. After all, in *Mṛṇālinī* the paradox was less acute. Paśupati's desire for Manoramā is not fulfilled in reality; they are united and separated in the same tragic moment at Paśupati's funeral.[34] In *Indirā* the plot is much more ingenious; both relations exist in reality. So it is no longer a matter of desire and fantasy; it is a fact of the world. In interior terms the two stories are of course wholly different. They show the novelist in very different creative moods. *Indirā* is lightly told, humorous, despite the desperate situation of the heroine, immersed in the comic trivialities of upper middle class life. *Mṛṇālinī* by contrast has a Gothic and romantic design, an inner sense of balance between the two different destinies of two of its stories—the tragic end of one set against the union of another. Manoramā is the direct opposite of Indirā. Indirā is urban, eventually reunited with her husband, telling a comic tale. Manoramā is an inscrutable figure of romantic conception, a dark tale with a dark and tragic ending in a world full of apocalyptic symbols—crumbling edifices, a temple on fire, of smoke swirling to eclipse a disappearing world, a physical obliteration which symbolizes a deeper historical obliteration.[35] Still, despite these differences between darkness and light, the comic and the tragic, there is an undeniable underlying philosophic unity. This is the unity of a question to which the novelist is trying to find an answer by varying contexts and fictional bearers. Both women know of their marriage, but use it differently—Manoramā to raise a barrier against Paśupati, Indirā to sanction her transgression.

The distinctness of the physical and the social relations is emphasized precisely by their indistinguishability. In Paśupati's case, the object of desire eludes him; Upendra has greater luck with his immorality. He lives in sin with his own wife. Is that possible? Is it simply a weak plot in which its unrealism escaped the author? Is it simply a weird construction which classical writers would sometimes use to show their technical virtuosity? Is it a philosophic problem about identity? Is it a life of sin for Upendra and not for his wife? Was she right in inducing her husband to sin? Is not making others sinful itself sinful?

Every moral code is known by its essential distinctions, definitions of opposition, by means of which, by elaborating which, it maps the entire space of its moral world. These are the axioms of its geometry; the standards of its measure. Clearly, the basic standards of judging acts and relationships are set up by the dichotomy between the sanctioned and the unsanctioned. The physical substratum is the same in both cases; but the mixing of social acceptance with desire makes the marriage tie sacred, its absence a violation. The security of the social world requires that these two contraries are held in place, that their limits, their boundaries, are marked with absolute clarity.

It is around the area of these limits that Bankim's artistic curiosity plays. His characters are people who are nearly always living their lives close to these regions of intersection, of liminality, where opposites come to play with each other. If there is a central theme in his novels, I suggest it is this enquiry, this concern about the nature of the liminal. The subject of his literary imagination is this endlessly renewed and endlessly indecisive conflict between the turbulence of desire and the stability of social norms.

At crucial points of the story, desire bursts through the barriers of social interdictions to create dramatic tension. These transgressions vary a great deal—from the subtle and wistful, often unnoticed by the characters themselves, particularly by his dimwitted heroes. It begins with a momentary look of longing on Āyeśā's face which betrays her emotions to her remarkably unsubtle prisoner. One thinks of the kisses bestowed, at crucial moments of the narratives on Praphulla and Rohinī. In many others the heroines escape such agreeable defilement very narrowly, and certainly not due to any moral exertion on their part. Indeed, they lead their men into these indiscretions with a coyness which can be seen as ontological.

It can be argued that Bankim never gave up this puzzling over personal relations, and this teasing possibility of several identities being indiscernibly superimposed. There can be an interesting comparison between *Indirā* and *Sītārām*, his last novel.[36] The moral problem in *Sītārām* is remarkably similar to *Indirā*. A wedded wife, renounced in early life, comes back to Sītārām, her elemental attraction swamping the barriers of interdictions imposed by his family, or expectations of propriety among his subjects. From a purely formal point of view, Bankim's experimentalism has grown more daring. In the Hindu moral order, her sexual role is central to the woman's identity; and in this the role of wife is the status of normalcy, for in this case, the passions follow and obey the social map. At the same time, the role of wife has contrasts on two sides, not just one. In his earlier novels Bankim returns obsessively to the more ordinary contrast—between the woman of normalcy and the woman of transgression—dichotomizing them, collapsing them, merging them into curious forms of indefinability. Now he tries something more complex—interweaving *three* roles. The role of wife is taken as the standard, the norm; that of mistress is related to this as a *lower* case of liminality, lower because morally condemnable. It is lower also, perhaps, because the woman who gives herself outside of social sanction gives priority to passion over the rational necessity for self-control, or mortification of desire. Since in the Hindu moral code high and low is often measured by this gauge—of access or renunciation of desire—which yields the distinction between the *gṛhī* and the *sannyāsī*, it must also do so in the case of a woman, between a woman who has access to desire and a woman who renounces desire on her own—a considerably more remarkable feat because she is supposed to be the object, the creator and therefore the symbol of desire. We shall see later in what exact sense she symbolizes desire—not as might be wrongly believed as a fault in her individual personality, but as part of the world's design. This offers a more interesting and complex possibility of constructing an identity problem. Alongside the normalcy of wife, we now have two types of liminality—a woman who renounces desire and crosses a limit above, and a woman who falls for it and crosses a limit below.

Consider now the different aspects of Sītārām's relation with Srī. At the initial equilibrium, she begins by being Sītārām's lawfully wedded wife, the relationship of normalcy, of uneventfulness; but at a time of life when it could not be seen as a normal relation of sexuality. Afterwards, as in so many of Bankim's stories, there is obligatory interference by a fraudulent version of fate. Sītārām's family finds, on having her horoscope read (and the reader makes a true statement of truly oracular opacity) that she is fated to cause death to her dearest. It is a statement which is to be interpreted twice,[37] and the nearer interpretation is characteristically wrong. She is estranged from Sītārām as a consequence. They have an accidental second meeting which revives their relation. In fact, Sītārām's conduct on this second meeting is not above reproach; for he wishes to take Srī back not out of a moral conviction of injustice done against her, or of her rights as his wife, but for the very different claims of her beauty. Sītārām gets attracted not to his wife—a social relation—but to the beautiful woman, a relation of natural instinct. For a spell, in a foreshortened sort of way, Srī is Indirā to Sītārām, but this is not drawn out, for that is not the point of *this* particular story. It has greater ambitions of constructing relational complexity. Srī takes a further step by renouncing her contact with her husband, becoming a *sannyāsī*, a renouncer of the world and its ordinary desires. When she returns to Sītārām it is with a new identity, she agrees to live near him without public rites of marriage; for the higher and the lower liminality are identical on one point: to both the mistress and the renouncer the path of marriage is barred. Sītārām's behaviour towards Srī is just what one would expect towards a mistress, one bound by this strongest of all relations. He is shown to be besotted. Srī is his addiction as no wife, no woman of normalcy can be. This can happen only with a woman who has the darkness of liminality about her, a woman transfigured by transgression. At the midpoint of the story, Srī has three different female identities superimposed on one another, which relate to Sītārām in three distinct ways, yet all subsisting within herself. She covers the realm of normalcy and the two liminal realms at the same time. She is a *sannyāsī*, one who has given up the rights of desire; she is a wife, a woman in the normal space; she is also perilously close to being a mistress, a woman who represents this dark sexuality of transgression. Obviously the question of transgression, of a fall, is exacerbated by the fact that it is not the fall of a housewife, but of a renouncer. It is seen exactly this way by the two rather prudish subjects of Sītārām's kingdom who act as a kind of chorus to their fate.[38] Srī is seen as going out of the social space of normalcy in two different directions. If relations define identity, who is she? What is she to Sītārām? What could be a good moral judgement on what she has done with her life? There can hardly be any greater indefinability than in this case. In a sense, here Bankim had achieved something like a liminal case of constructing liminality; the play with limits has reached a limit of its own.

* * *

In this world that we see building up gradually through succeeding novels of desire there are some underlying regularities of structure and form. The point of underlyingness, or implicitness is important. An artist as successful, prolific and creative as Bankim obviously followed deliberate plans and strategies of narrative construction. His fiction also shows a remarkable variety of situations, of personalities, of plot construction, a deliberate and very well-judged variety of language. Still, there are some underlying themes and forms common to practically all his novels. If this had been a real world, these would have constituted its structure and its laws of motion. There is of course nothing contradictory in this. What the artist depicts is the variety of a world, and to be a world it must have form and order. What is revealed by these is the author's sense of the world; what he thinks the world is.

One of the most significant features of this world, the way the world is, is what women do in its general scheme,[39] though perhaps the proper verb for what I wish to convey is not *do*, but *be*. Bankim must have been conscious of this ambiguous centrality of the woman in the transactions of the world of his making. At the most obvious level, this is announced in the names of novels after their central figures: *Durgeśnandinī Mṛṇālinī, Indirā, Rādhārānī, Rajanī, Devī Caudhurānī, Kapālkundalā*. This consists in a declaration by the author that these narratives are centred around these characters. Only *Candrasekhar, Rājsinha* and *Sītārām* bear the names of their heroes. But the logic of history and the logic of narrative are different. There is a sense in which even these stories which do not formally revolve round the fates of women, do depend on them for the movement of the narrative. Women are the prime movers of the verbs. Clearly, the story in *Candrasekhar* is as much Śaibalinī's story as Candrasekhar's, in fact, more so. For although Candrasekhar and Śaibalinī constitute one of the few pairs in which the relation between passive and weak men and dominant and active women has been formally reversed—because Śaibalinī has no remarkable trait except her beauty, the root of all trouble—still, it is she who is critically significant for the motion of the narrative. It is her life and playfulness which bring the disparate elements of the story together. It is the relation of three men with her—Candrasekhar bound by the tie of marriage, Foster by the tie of violence, and Pratāp by the wistful relationship of renunciation—which keeps the story going. She is the cause in the story of all the trouble that befalls everybody around, the great dispenser of unmerited suffering. Particularly poignant from Bankim's point of view is the relation with Pratāp—of a typical transgressive desire, which has no kinder solution than the finality of death. To anticipate, the tragic nature of finitude is particularly evident in the last paragraph of *Candrasekhar*, an explicit declaration of the problematicity of the world.[40] It shows what *Durgeśnandinī* began by showing: that it is wrong that we have to declare some things as wrong. There is a peculiar purity, innocence, nobility in the desire that Pratāp and Śaibalinī feel for each other; and there is something grotesque in the artificial purity imposed on them in their everyday conjugality with Candrasekhar and Sundari.[41] This again plays the game of shuffling definitions—making the right wrong and the wrong right.

In *Rājsinha* too, it is Cancalkumārī's decision to deviate from what her world expected her to do that creates the initial narrative movement. Rājsinha is certainly more significant than Cancalkumārī in terms of history, or of status, as a king, as a warrior, as her husband. But there is, as it were, an internal hierarchy of the interior world of the story and in this world she is far more significant than Rājsinha. She begins the initial events, she shapes it by her crucial acts; Rājsinha is in fact reduced to mere adjustment to the sequence of events begun by her initiative. Srī, similarly sets the story going in *Sītārām*, by doing an extraordinary deed. So without her, Sītārām surely would have existed as he did exist in history; but the Sītārām of the story could not have. He is literally brought into existence by her. In *Devī Caudhurānī*, it is Praphulla's ability to do the unexpected which creates the structure of the narrative. Curiously, it is always women who act extraordinarily, who exercise the frightening acts of choice. Women are therefore symbols of transgression in a double sense. They create or induce transgression by being objects or symbols of desire. But they are also literally the agents of transgressive or liminal action. They are seen as being peculiarly capable of breaking the world's moulds.

In both the early Lukacs and Girard, there is a theory that the novel form revolves around a problematic hero, because of an original lack of fit between their inclination, their truth, and the structure of the world around them, the truth of the world, between a subjectivity that is expansive and a world that is too narrow. In Bankim's world, interestingly, it is women who are always the symbols and vehicles of this problematicity. It is their internal conflicts or their conflicts with this narrowing world which make the story. Male figures, even when artistically forceful, are static and uncontradictory, too predictable and therefore unable to generate narrative tension. Jagatsinha's character is not weak in the ordinary sense, though those of Govindalāl, Nagendra and Upendra certainly are. Rājsinha, Sītārām and Pratāp are not portrayed as feeble men; yet they do not move their stories. This may have something to do with the identification of women with instinct which makes their behaviour simultaneously less predictable and more final, while men representing reason,[42] are fixed and unsurprising, and when acting, hesitant.

Bankim is not interrogating an everyday relation between men and women. As a novelist, he does depict such relations quite often, because a world full of extraordinary relations would be less than credible; indeed, they would not show themselves as extraordinary, and consequently would not be spectacular. But these ordinary relations are a part of his field of vision, not its object. They show unmistakable marks of fictional indifference. They are not what interests him, what he is interrogating; the object of his search is the strange, the abnormal, what I have called the liminal.

Marriage relations may be made in heaven; but they are, with frightening regularity broken on earth.[43] Still, though women draw men towards them, they do so with a curious innocence. We must understand the nature of this fatality which women carry within themselves. They play a fatal role all round. It is not only men whom they drag to destruction but also themselves. Nearly always they strike as a force of uncomprehending destruction. They do not do it out of design, and it is therefore not attributable to some moral flaw in the person who plays the role of fate. That women, especially attractive ones, are destructive of families, of the straight and narrow path, is not due to their personal malevolence, weakness or baseness of character. It is part of their being. To accuse them for being so is as sensible as blaming a bird for flying or a flower for being beautiful.[44]

Playing variations between two stories with one central problem, is as we have noticed, part of Bankim's fictive style. For instance, *Kṛṣṇakānter Will* is a play on the logical obverse of *Viśavṛkṣa*. Bankim's social conservatism is so well documented that this comes out in an even more curious light. He depicts the destruction of domestic peace and harmony with a marvellous detachment. However, this attitude does not reflect sympathy for transgressions in the ordinary sense of the word. There is a peculiar descriptive finality about them. It is not the attitude of someone who approves or condemns such behaviour, but of one who believes that these are beyond such judgements. Undeniably, there are foul characters in Bankim's novels. But what is remarkable is that what both evil and innocent characters bring about by their actions is nearly the same. Rohinī and Kundanandinī illustrate this identity of consequences. As individual characters, they are opposite to each other. Rohinī is,[45] at least after she plans her crime, close to a figure of evil. Kundanandinī, by contrast, is the picture of innocence and simplicity, entirely devoid of ill will towards anyone. But the consequences for the two families which come in touch with them are identical. Even if this is not a deliberate construction, it makes a point with particular clarity. Differences in character are not significant for the determination of fate. It is not their intention or their personality which cause destruction, but their very being. Typically, in nearly all of Bankim's novels of crime and punishment, there are indications pointing towards either a philosophical or circumstantial extenuation. Often, the character who transgresses, usually a woman, at a time of crisis gives a powerful and entirely credible justification of what she has done. Often, the narrator shows how she could not have acted otherwise—a questioning of the concepts of guilt, responsibility and determinism lies at the heart of Bankim's art. It is in this sense that his art, when he aims at it, attains the truly tragic. Like all great tragedies we seem to encounter in them something that cannot be reckoned in terms of human agency and responsibility alone. It is what happens inevitably and defies accounting for in ordinary terms of cause and responsibility that is called fate.

Could it be objected perhaps that Bankim could not have thought of liminality and transgression as truly tragic, and this is too Nietzschean, at least western, to be true about him?[46] Could he have thought of this as part of an indigenous tradition? I think he could have derived this way of interrogating liminality and understanding the architecture of the world from the classical Hindu tradition, or at least a certain reading of it. Surely, not all Hindu myths have this kind of design. The *Rāmāyana* for instance is a classically simple story. It is an epic in simplicity, the simplicity of the story, of its emotions, its fundamental conflict, its eventual resolution, and the utter undisturbed symmetry of its moral universe. Perhaps one interesting fact about it is how so simple a story could be poetically so compelling. Its figures show a highly stylized idea of the finality of the simple. The ideality of its universe is another; and such ideality may be considered necessary in the daily struggle with the untidiness of real life. At the end of the day, as a foil to the complexity of experience, it is reassuring to represent this fundamental map of the simple relations which constitute the structure of life. It is natural therefore that it is the *Rāmāyana* that would be part of such daily rituals of iterating how the universe is built.

By contrast, the *Mahābhārata* seems more concerned with liminality, and Bankim has an endless curiosity about it, in contrast to his general indifference to the *Rāmāyana*. Either the *Mahābhārata* story, or some of its derivative legends, like the folk mythology of Kṛṣṇa, arrange careful encounters between the settled norms of everyday life and an experience of the liminal. If one makes an inventory of the roles by which a common Hindu is expected to live his life, the *Mahābhārata* and the related complex of myths could provide interesting counterpoints to nearly each one of them. Women must not bear children outside marriage; but Kuntī does, without much apparent defilement. The career of this child also belies the possible belief that such children are miserable products of sin and destined to a life of sorrow. Despite a contradictory fate, he is a child destined for glory, to being exalted, though he bears the mark of transgression on himself.[48] Draupadi's marriage to five brothers puts even Bankim's argumentative ingenuity to considerable strain. Similarly, there are extreme interdictions against sexual relations with women who are older, married, particularly to one's relatives. But the story of Kṛṣṇa's love for Rādhā transgresses all such norms. Clearly, the purpose of such stories is not to turn people away by their wantonness, or to show the inevitable ugliness of life.[49] Transgression here is exalted by the touch of the divine, so revulsion could hardly be their point. Often, there is a prurient delight at the exploits of the divine transgressors.[50] After a public exultation over clinical accounts of erotic play between the divine but unaccredited lovers,[51] members of religious congregations trudge back to their severely ruled lives, in which even a mention, a mere thought of such abandon would be sacrilege. Two extenuating interpretations are usually offered to mitigate their wantonness. A common though hardly convincing answer is that this is to show the difference between humanity and divinity. Another construction is that these are a metaphorical taking away of metaphorical clothes to reveal the metaphorical purity of the devotees' souls.[52] But there are difficulties about such explanations. It is a curious way of making men obey sexual mores to show that gods rarely observe them. Besides, in cases like the *Gītagovinda* metaphorical joys are depicted, and could be received with a suspiciously unmetaphorical delight in their materialism;[53] and secondly, other metaphors could perhaps have conveyed such conceptual ideas equally well without causing the carnal delusions which may arise in this case. A third different construal of this kind of thing would be that this shows a concern for the problem of liminality, though there can obviously be varying interpretations of its possible reasons and cultural implications.

If Hindu scriptures show a concern for liminality, this could furnish a possible source for Bankim's interrogation within a tradition he so obviously valued. It is not pleasing or prurient, but morally significant and philosophically important to come to the edge of the world and peer at the darkness beyond, to stand at the limit and look at the other side. That alone can create a sense of what it is a limit of: it is an understanding of the other that creates a deeper sense of the self. An interrogation of moral limits is quite different from the easy moral nominalism which believes that it is simply necessary to name something a moral guilt to begin its correction, to produce a simple unfailing recipe for virtue. If there is a doctrine of virtue in the *Mahābhārata*, that seems considerably more complex—showing how morals tend to break down in liminal cases, how people often have to invent a morality, i.e., ways of acting uprightly and well, for unprecedented situations. To live a good life is not an uncritical pursuit of immutable rules which can guide behaviour in all conceivable situations. It is a critical adventure. Philosophically, the interrogation of this moral other is similar to the great question of the other of life, death. Death is not something one can name familiarly, something that can be described like other experiences; it is, as the *Tractatus* put it, not an event in life.[54] Death in that sense constitutes a limit to the discourse about life. A language which is at home in life, when confronted with death, cannot pass a judgement, because it does not know what kind of judgement to pass. It cannot describe because it faces an occasion when describing is somehow not the right thing to do. Death extracts this kind of philosophic seriousness, a respect for the unknown, the liminal. The *Mahābhārata* treats moral liminality the same way—not to treat it simply and pejoratively in terms of everyday morals, but to respect it as a genuine limit, an occasion when the everyday discourse of morals falls silent. Sometimes, these are ascribed to the divine; i.e. to signalize, to mark symbolically that the discourse of man stops here; here we require a different discourse. To understand what discourse can bring into speech, it is to be taken to a point where discourse must fall silent.

It is not a long way to travel from this quest for the liminal to the aesthetic form of the tragic. Liminality can evoke two responses: an easy response of defining it into something akin to criminal, or a more complex response respecting it as a form of the unknown. The second kind of mind would consider the liminal—in the cognitive, the moral, the living—as the only thing worth seriously exploring, or worthy of wonder. This affiliates the liminal to the tragic; for an enquiry into such liminality, tragedy is the only proper form. And Bankimchandra's artistic world, despite its occasional transcendence through laughter, is essentially, ineradicably, tragic. Such a world would not comfort an ordinary conservative. In this world, luminously clear, uninfringable relationships are bent out of shape, sent into a mysterious abeyance. It does to the relations with which we constitute the social landscape of our everyday world exactly what romantic painting does to our feeling of the invariance and availability of nature. It grows strange, alien, mysterious. The darkness of the world is filled with the debris of definitions we take for granted. It is, in that sense, a typically romantic moral landscape, a world of beautiful and sublime destruction. And this world finds its proper representatives in the beautiful ladies of darkness.

There is an idea commonly applied to painting which could perhaps be used to understand the darkness of Bankim's world. Such perception is not always a matter of deliberate choice; they do not really perceive the world that way because they choose to. That represents a certain distillation of their experience of the world. Even though they sometimes wish to say something quite different, this perception comes through in what they wish to say: through what they are trying to say, something else gets said, something which is also theirs, but in a more profound, non-deliberate way. Such views, art critics say, come out sometimes in painters' use of colours. As time passes, an artist's view of the world may get darker. Often the same objects are depicted, after twenty years, in an indefinably darker tone. What is smeared on these objects of the world is not just paint, but something more fundamental to painting than paint itself; they are smeared with the artist's sadness. It is in this sense that Rembrandt lived in a world that is dark. As he goes along, Michelangelo's palette and his world get darker and more mysterious.

In literature too artists tend to respond to the world in their own individual ways, and depict it in their tones. To use the tone in case of language rather than colour is not wholly inappropriate. Frege uses the term tone to indicate those qualities a proposition may have other than its meaning, i.e., the quality about which one cannot say that it is either right or wrong. This does not mean that the artist will not wish sometimes to depict the world in a different mood. He must surely do that if he is a real artist, if only to see how the world looks like when it is not what it really is. Thus we have radiant and lighted paintings by painters who specialize in creating the most threatening of nights, and whose genius is at home there, who feel as if the day is somehow discordant, is somehow merely an interlude in the night that represents the nature of this world. This metaphysical tone in Bankim's art seems to me to be predominantly tragic. His stories show this darkness in the peculiar fragility of happiness, the constant lapse of smiling, inconsequential lives into unmerited and uncomprehended suffering. It reminds me of one of the damned in Michelangelo's 'Last Judgement'; what is arrested in his face in the final moment of his metaphysical existence is a rather peculiar emotion. It is not a face of contrition. It is not a face of ordinary terror, nor the face of one who sinned boldly and is now cowering before his maker. It is a face expressing a dark and terrible wonder, of a creature startled by the smallness of the transgression and the eternity of its retribution.

In this tragic world happiness, rationality, the human ability to create an order of whatever kind, a moral order, a rational order, or the small order of a little family of ordered unthreatened relationships, is forever threatened by the storm of the universe, the night—an incomprehensible and irresistible force which can destroy everything. Stories of human life are allegories of the world; they show the utter contingency of everything. The night we are talking about is the night of the world. The only conceivable attitude towards it is of wonder. For this is no ordinary night, not a night which passes obediently into a following day. It is a night that resides in the nature of things. Bankim's world is also nocturnal, with the darkness of a night without end.

Art, it is said, is a rhetoric on the world. An artist experiences life in a particular mode, and decides through some largely inexplicable process that it is in the nature of the existence of men to be in a particular way. This determines the tone of his art. His art may reveal a world which is dark, mysterious, threatening, nocturnal, like Rembrandt's. Or the world of a Rabelais—of sensuousness, mirth, full-blooded enjoyment. It can be the world of the Classicist painters—orderly, measured, rational, perpetually bathed in a rational sunlight, throwing orderly shadows, where even disorder leaves enough reminiscences of order. Or it can be the world of the Romantics' landscapes—where the smallish earth is dwarfed by a majestic sky, in which the world is being torn apart by an unending storm, trees bent in the wind, where everything is at the edge of this darkness of a night which will not end next morning, but a metaphysical night that will never lift.

Most of Bankim's better novels have a metaphysical air of the tragic. They may not in a formal sense always bear a tragic form. What does the tragic mean in Bankim? Of course there is a lot of levity in his prose, and in his narrative construction. Tragedy here nonetheless exists like the tone of an artist's painting. That seems to reach the essence of life's meaning for him. He is not always deliberately tragic; rather this sense of tragedy transcends the strategies of his deliberate craft. Some fundamental beliefs of people colour all that they see without being present in their conscious design, or indeed in their own consciousness. For seeing itself is not an object of seeing.

Tragedies can arise through two different modalities. There are what can be called tragedies of suboptimality, of faults or flaws of character. Individuals can tempt fate by their lack of self-control, or their predominant passion. In the tragedies of Corneille for instance, this occurs through an excess of a sense of glory.[55] People can tempt fate by not checking themselves, sometimes by not knowing themselves. They meet disaster through an excess of ambition, power, covetousness, caution, even love. Such errors are faults of character; we see the personalities as flawed, one-sided. Often they take decisions they need not have taken. In these cases an equal sense of finality does not arise because it is easy to conceive of an alternative, a case in which the mistake is not made. There are tragedies of a second kind where the fatality arises not out of an error, but from finiteness. They arise from the fact that people cannot be everything at the same time, or the fact of being just what they are, for no fault of theirs. It is a pity that Jagatsinha must choose between Āyeśā and Tilottamā, that he met both of them, and not one; that Kunda cannot buy happiness with her innocence; that Pratāp cannot love Śaibalinī and cannot stop loving her. These are tragedies of choice, where the tragic thing is not the choice made, but the obligation to choose. It is the structure of the world which requires these choices, that there is one at the expense of the other.

Tragedies are in this way close to aporia—questions which do not admit of answers. Or they spring from the conflict between two equally inevitable things. Underlying all of Bankim's novels is a recognition of this kind of conflict—which tears people apart, which has no solution. As a *conservative* he finds the formal necessity of social order immutable and inviolable; as an *artist* he finds the surges of desire equally immutable and ungovernable by human conventions. Marx offers in his comments on Balzac a fertile suggestion about the nature of writing. If his view is sensible, then there are two writings in Balzac or two visions. Balzac sees the world and writes as a legitimist: this is the author's vision; but there is also a vision of the text which unravels capitalist social relations as no one else could in that generation. Balzac did not intend to do this; yet Balzac does do it; and it is odd, to say the least, to assert that his depiction of capitalism is entirely unintended. Obviously this calls for a more complex theory of textual intentionality, a theory of seeing in terms of degrees what it is normal to see in terms of a simple opposition of the intended/unintended. Anyway, the truth of Balzac's texts would have to be something complex, which cannot deny the truth of either of these two visions. The truth of Bankim's consciousness, I suggest, has to be similarly constructed. A proper slogan for his art would be that not marriages but transgressions are made in heaven, a puzzling idea indeed for a Hindu conservative. As both of these are inevitable and permanently at work, the world moves permanently in conflict; this is the reason for the existence of art whose object is not just entertainment but also philosophic wonder. Individuals in that sense are mere instruments, a character's central position in a work means simply that 'he is particularly well suited to reveal something problematic in life'.[56] Bankim's art creates life stories of individuals to express wonder at this primeval darkness, this peculiar problematicity of life that is much easier to think of through art than through conceptual thought. In *Kamalākānta* this wonder finds explicit expression, in a passage about happiness which is not a happiness of enjoyment but the startling happiness of wonder: 'You have no one of your own in this world, only your happiness; I too have no one to call my own, except my happiness. . . . I call the beautiful. I call the good. I call whoever is listening to my call. This wondrous universe that stands in front of me which I cannot understand, only wonder at, I call this universe'.[57]

A world is held up and stays in place because of such limits. Consequently, there could be nothing more important for someone wishing to know the structure of this universe than to know these limits: because to know the limits is to understand the order of a world. To know a limit is to know the two things it separates and gives identity to; it is these which structure the world and give it form. If we cannot really know these limits and definitions, we know at least their fallible and human character, their provisionality, their constructedness. If we cannot classify and judge all human acts and all human fates by the small repertoire of judgements that a moral code provides, through this cognitive failure we learn something far more important. We accept the provisionality of our happiness. We learn to forgive. We grasp the secret of kindness.

2

Humour and the Prison of Reality

Kamalākānta as the Secret Autobiography of Bankimchandra Chattopadhyay

Nothing in discourse can be so many things, carry so many meanings as humour. Perhaps nothing else in discourse is so completely free of the usual demands of linearity. Bankimchandra's *Kamalākānta*[1] shows in Bengali literature the spectacle of the manysidedness of laughter, its indefinability. My text here is less a book than a manner of writing. *Kamalākānta* is not the only place to look for Bankim's humour. Even when he is not writing explicit satire, he tends to take refuge in it. He made it his characteristic weapon, his way of attacking without being attacked, criticizing those with whom he does not wish to enter into a dialogue. Humorous writing is in any case an unanswerable weapon, and few of his generation could use it with comparable effect, few indeed in the entire history of Bengali literature. *Kamalākānta* shows the great versatility of a literature of laughter, its internal repertory; it is at different times the laughter of contempt, of not taking something seriously, the laughter of reproach, of pity, the laughter about the self which feels it is part of what it is laughing about. *Kamalākānta* is the greatest example in Bengali literature of Bakhtin's argument about laughter, its having a key to the world's mysteries which other forms of more serious literary writing lack:

> Laughter purifies from dogmatism, from the intolerant and the petrified; it liberates from fanaticism and pedantry, from fear and intimidation, from didacticism, naivete and illusion, from the single meaning, the single level, from sentimentality. Laughter does not permit seriousness to atrophy and to be torn away from the one being, for ever incomplete. It restores this ambivalent wholeness.[2]

The fundamental secret of the power of humour was wonderfully revealed in two lines in Sukumar Ray's 'Khicudi', the poem which reports that the duck and the porcupine carried on a traditional existence until a time when, by the miracle of humour, the 'grammar of reality' fell away, and they formed a mixed animal.[3] Humour precisely allows us to break the grammar of reality and history. Kamalākānta, the protagonist of Bankim's humorous essays, is an ungrammatical being, in this sense, half conformist, half rebel.

Laughter is an equivalent of play in art. Art, it is often argued, is play compared to the seriousness and deadliness of everyday life: the beauty of art a foil to its drabness, art's poignancy to its tendency to slide into meaninglessness, art's clear form a foil to its endless and shapeless stretch. Even within art, which can be serious, humour has a special place as play. Compared to more formal literary styles, it is playful. That is why in humorous art we find a peculiar gaiety of liberation from encumbrances of form. There is a peculiar openness of form in *Kamalākānta*. It is in part serious ironical prose, in part essay, story, drama mixed together. In terms of writing, it takes the most unpredictable of turns. As it is non-serious, it is free from obligations of consistency, it can play between its own possible meanings. It can be any of them, it can be all; it can be, precisely because it is laughter, this irresponsible play of astonishment.

The text we are going to read is a text of madness. Its putative author, Bankim says, was regarded by ordinary people, as mad.[4] The madness that produces this text is, however, a text in itself, a sign. It can be read, made to yield its meaning. We must first understand *who* says this text, before we can grasp *what* it says.

In a manner of speaking, the foremost question is about understanding the almost unknown author of one of the most celebrated texts of Bengali prose. This is written allegedly by Kamalākānta Cakravartī, a Brahmin, homeless, occupationless, classless,[5] a drug addict, a parasite, a sayer of the unsayable. Curiously, very little is usually said about this remarkably odd figure. He is wholly merged into the authorial personality of Bankimchandra. But he is not so negligible; because he exemplifies, to my understanding, not only an important segment of Bankim's art, but also his strategy of intellectual existence. Still, it is hardly ever asked in literature why the author creates another author who is such an interesting negation of himself. What does the author's alienation of his own text signify?

On his own part Bankim clearly took great pains to put the text at a distance from himself, alienating Kamalākānta's utterances from his own by a series of symbolic disjunctions.[6] Bankim is sane, Kamalākānta is suspected of insanity; Bankim is staidly respectable, Kamalākānta is marginal, of doubtful (unspecified) occupation; the author normal, Kamalākānta an opium-eater; Bankim is a solid civil servant, a salaried man of a decidedly upper bracket; he is dependent on something of a mixture between respect and pity; Bankim is a realist, he is a dreamer. To state the argument in more serious form, Kamalākānta is a bundle of *negative* attributes, indescribable in terms of the available categories of social description.[7] Whatever he might be, he is, quintessentially, *not* a *babu*. Bankimchandra was a model of the successful Bengali gentleman (bhadralok; the English term 'gentleman' is a common but strictly inappropriate translation) under the colonial social dispensation—successful, contented, reconciled. Kamalākānta is his exact opposite—idle, disorderly, unmarried, unkempt, socially indescribable, an irrationalist, or perhaps, a rationalist dedicated to the cause of what his contemporaries would have seen as 'irrationality'. Yet, despite his incongruity in this babu world of nineteenth-century Bengal, Kamalākānta is something undeniable. He is everything that Bankim is not, or in a different reading, everything that Bankim would like to be in his dreams of an alternate world. He is his self which has broken the syntax of reality, his self in a dream. He is, in a manner of speaking, writing the secret autobiography of Bankimchandra Chattopadhyay.

Indeed, *Kamalākānta* cannot be read adequately without decoding all its pretences. Some of these are of course fairly obvious. It pursues a line of thought entirely unrepresentative in this Bengali 'age of reason'. It is quick to remark that in colonialism everything that was part of European history tended to have a parodied reenactment. It refers constantly to the effort at reenactment of European history by the babu and destroys it by satire. Parody contains a principle of this alternative construction of reality. By a change of pronunciation, an altered inflection of the voice, by changing a word, parody achieves a startling displacement of meaning. Bankim's humour uses the myths, the tropes, the hierarchies of significance of Enlightenment discourse itself to undermine that discourse. His special favourite is a form of humour Sanskrit poetics called the *vyājastuti* (literally, false praise). In that poetics all literary expression is called *vakrokti* (utterances which have some obliqueness about them); but this has an obliqueness that makes it *vakrokti par excellence*. And there is an adequacy between the obliqueness of this form and the obliqueness of Kamalākānta's vision of the world. The world, especially the social world, hides its truth; it requires an oblique, a second sight to see it.

Defying the grammar of reality has some implications which ought to be clearly stated. Humour has a more open structure than other types of discourse because it regards more possibilities as admissible than real life does. That is why humour is critical and subversive, and particularly effective against those assumptions of social life which pretend to some kind of self-evidence. It is hopeful precisely because in ridiculing the reality of social relations it asserts the important fact that the world that exists is always short of, is a travesty in a sense, of the world that ought to be. Other types of thinking do this too. One obvious parallel would be religious utopian paintings of paradise, where non-violent lions live on excellent terms with deer. Another could be the utopia of a revolution which defies a grammar of a political kind. Conceptions of revolution, of paradise and humour—all, on this view, would have something in common—they all defy the colligation of reality, they all have the gift of seeing as historical and contingent what to ordinary eyes appear as necessary.

A study of Bankim's humour must be an analysis of its totality, not just a part. I would not like to associate his humour too closely only with the mood of defiance, for there is also a sorrowful, dark aspect to it—the fact of its mocking itself in the act of what it says, its perpetual tension between the babuness of his existence and his intellectual rejection of the babu social and ideological order, and what could be called an implicit rejection of this rejection by choosing to do it within a subterfuge. This indicates, at least partly, a bad conscience, not an unambiguous attitude of rebellion. It is followed by a certain ineradicable sense of guilt. Obviously, he was not able to attempt what, if radical historiography is to be trusted, the unlettered peasantry undertook at times. Bankim in that sense retreats from the more demanding task of 'changing the world' into merely offering an interpretation. The babus get his contempt because they translate acts into discourse, which represents movement in the wrong direction, turning away from the obvious practical syllogism of colonial political existence. Instead of going forward from liberal, rational premises into political action the babu collapsed back into discourse, substituting radical propositions for radical acts. Some of the humour, particularly its occasional pessimism, must be linked to the perception that Bankim himself could be rebuked for doing the same thing, and the limits of the existence of the babu that he analysed with such power turned out to be the limits of his own self. Irony thus turns back upon itself; it can be seen in an ironical light. It is common knowledge that insanity is a way of living with unresolved contradictions. This could be one reason why the speaker of this ironical discourse uses the literary sign of madness. But madness means different things in the cultures of rationalist Europe and traditional India. I wish to suggest that the madness out of which Kamalākānta speaks is a complex sign constructed out of elements taken from both these cultures, giving him the privilege of an unprecedented insanity.

In quite a literal sense, Kamalākānta offers us a play of two sights. He is gifted with a second sight which opens up when he takes opium. To the ordinary sight the world is closed in its transparence. It can see only self-evident things. Opium intensifies his sight, and when he takes it, what he sees is a world of essences alone. To this sight beings turn into objects, objects into beings, cats begin to speak and carry on socialistic disputation of a delightful and convincing irreverence. In a word, the world opens itself, and yields its secrets to this second sight. Take the cat's discourse on property, his inverted socialistic vision of the world. It is a rationally defensible construction of the social world, but it does not figure in the discourses of the real world, because there is no one to say it, the contingent absence of a speaker of true speech. But truth should not be voiceless simply because it does not have a speaker. In a world of essences, it can be spoken by a political cat.

Since it reveals essences, something that ordinary sight cannot, people need this second sight, which shows the structure of the world as it really is, quite different from the way it appears to be. He too could say, echoing someone else, had the world of appearances been the same as the world of reality, human beings would not have needed opium. It is wrong to say that when he has opium he sees better, because that would be a physicalist idea; it assists him to see in an altogether new way. He sees the world through the logic of alternatives, as it could have been, as it could, alternatively, be.

Just as in Foucault's story, rationality defines madness in order to say what it thinks about it, in exactly inverse terms, in *Kamalākānta*, critical thought assumes this title of madness to say what it thinks about rationality. In this I see the working of both an internal and an external tradition which are opportunistically blended. The trace of the European definition of insanity is of course very clear. But we must also seek more seriously for the influences of an indigenous tradition of the reverse type to understand this unprecedented insanity of Kamalākānta. The liminal figure, or rather the outsider here is not entirely like the Western figure of the madman, he is much closer to the indigenous figure of the renouncer, the *sannyāsī.* When saying his farewell, Kamalākānta explicitly marks himself as a sannyāsī: Kamalākānta is a sannyāsī at heart' (*Kamalākānta antare antare sannyāsı*).[8] Unfortunately, he cannot be a classical but only a debased renouncer, because all possibilities are debased in his culture. He is both a sannyāsī and his caricature, just as Bankim is a rebel and a caricature of a rebel at the same time (*antare antare*). All that the babu inside Bankim is capable of is this secret resentment and a secret rebellion; such things like secret rebellions, are really dreams, or judged more severely, a defilement.

To act sensibly is simply to act rationally within certain ordinary acceptable premises, but these premises of everyday rationality are not uniquely grounded and justifiable. They indicate mere appropriateness, and its premises, it is readily conceded, would not hold under serious philosophical questioning. This theory of an inversion of the everyday sight in the truth perceived by a higher sight is simplified by popular consciousness into a strange equation of religious insight and madness. Anyway, this could in fact turn the rationalistic theory upside down by claiming that it is those who do not live within the limits of rationality who can speak the truth.

Implicit in this doctrine therefore there is a theory of multiple rationality. If rationality has two principal criteria—of being internally consistent, and of being in some sense consistent with the world such that it allows us to handle this world effectively—then, it is implied, there could be more than one way of acting rationally. Ordinary relations with the world, represented by the *gṛhī* are of course rational in an undeniable sense. But the other's way is also rational. What is remarkable is the relation set up between these two ways of acting on the (at least the social) world. Practitioners of the other rationality are tolerant of the ordinary register, they do not call that irrational. Most of these traditions assert the possibility of transfers, of passing back and forth, so that the reasonableness of one way of acting can be appreciated from inside the other. Thus there was available a living tradition of thought which celebrated marginality, and treated it as a sign and qualification for seeing great truths of existence.

At any rate, these reveal a certain *figure* of thought, which associates privileges of either cognition or of sentiment, with the *outside*. The right to the truth, as the right to love, belongs to one who has the courage to step into this great unknown outside.

Bankim's *Kamalākānta* belongs to this tradition, but with a certain obliqueness, both an ancestor and inheritor. He is an inheritor of this curse (or gift) of an inspired liminality, being able to see what others cannot. He emphasizes by signs the tragedy of prophecy, its mixed quality of being half gift and half curse. He is also an ancestor of Tagore's *bāuls* and *bairāgīs*. In both his and Tagore's works these outsiders are marked by a different language. In Tagore these characters most often leave their commentary on everyday life in songs—a rhythmic, harmonic counterpoint to the prose of daily existence. Kamalākānta too speaks a different language: he speaks in humour, in a different key from the humourless seriousness of the everyday language of interest.

Kamalākānta's insanity, I suggest, has something to do with the untidy, tortured, self-doubting beginnings of a critique. This is reflected in the patchy clarity of his utterance. In individual parts, Kamalākānta's arguments are clear, and theoretically sound; but it does not add up to a general solution to the questions that are latent in his discourse. He sees clearly that colonial representations of western rationalism need critiques, but not very clearly what that critique would be in its full form. And madness offers a wonderful sign of this indecisiveness, a subtle way of asserting statements, but putting these statements themselves under a sort of questioning at the same time. Madness gives this whole undertaking a philosophically tentative, interrogatory form, a sign of the unconcluded nature of this critique.

Humour, by itself, is an abstract idea, because it can assume various forms. Of the great number of possibilities, I classify Bankimchandra's humour in *Kamalākānta* as double irony. Double irony has a simple formal feature which distinguishes it from other kinds of humour. It not merely mocks at the world; it does also, in the same act, mock at its own mocking at the world, and therefore itself.

Does this sort of discourse when used in high literature, indicate anything beyond an individual predicament or a personal style? Does this show something interesting only about the author of these essays, or are there perhaps signs in this discourse of a shared, collective difficulty? Are there some people, some classes, some political groups, who must speak like this, in this manner of saying and not saying, hiding what they say, turning their whole discourse into an endless *double entendre*? It seems to me that this is the typical discourse of *subalternity*, of a subaltern group which feels the inevitability of criticism and the inevitability of its practical failure. Also quite typical of the subaltern is this tragic disingenuousness, of making criticism, and hiding it in the act. Subalternity is a tragic fate, and for those who are not able to come to terms with it cognitively and practically, the only recourse, the most ambiguous recourse of all, is the discourse of irony. Abstractly, the two kinds of discourse, the tragic and the ironical, could be very far apart; but it is this situation of historical subalternity which constantly causes them to form peculiarly unstable configurations, and Bankim's thought moves from a tragic perception of the world to an ironical way of dealing with it.

In mocking at the world, it can also mock at its own mocking at the world. This is not just a rhetorical phrase. This implies that if all that the humorous statements say about colonialism is true, then it could hardly do to merely mock at it; and we are back at the guilt about practical syllogism. In Bankim the critique of the babu by Kamalākānta is mixed with a critique of Kamalākānta. It comes close to a nihilistic pessimism; for it satirizes power but does not forget to satirize helplessness. It has that rarest of ability to turn the critique towards the self. In calling the world in question, it can also call in question its own pretensions. It is in that sense a deeply tragic sense of the comical. For the sense within a discourse that it is a comic spectacle is an ultimately tragic sense. Ultimately, it is an admission of defeat, of its inability to do anything to the world of injustice, leaving this world regretfully as it found it.

The Theatre of Justice

Humour, I take it, is a way of inverting the world, a way of escaping from what is undeniably and intolerably true. I shall begin my textual reading of *Kamalākānta* with a gesture of inversion. The text has three parts: the *daptar*, his diaries; his letters to the editor of *Bangadarśan*, and a last indescribable piece in which Kamalākānta takes on his favourite enemy, the judiciary, his eternal and favourite object of derision, called Kamalākānta's deposition (*Kamalākānter Jobānbandī*).

In this peculiarly compelling scene of the court the whole theatrical field is set out with dramatic clarity, so that we can, if we look closely, see what is involved. The event is formally a judicial process, in which there is a great theatre of dispensing a just verdict; but what it actually achieves in the end is a strange combination of solemnity, injustice and farce. The whole colonial dispensation is present there in a transparent code; it is a powerful condensation of all colonial relationships with a miraculous authenticity. Inside this arena there are two dramas going on at the same time. There is of course at the surface level a judicial case, with its plaintiff, defendant, pleaders and judges; the plaintiff is Prasanna, the milkwoman of Kamalākānta's particular acquaintance; the defendant a hazy under-determined Muslim (*neḍe*), the witness Kamalākānta, and we have the entire theatrical personae of *hākims*, *moktārs*, *muhurīs*. But it is easy to see that there is a second judgement being prepared; in this deeper case, the people who get judged to their detriment are Kamalākānta and Prasanna; and those who do the accusing and judging (crucially together) are the other set, from the judge down the line through the lawyers (on both sides, because both show an equal incomprehension of Kamalākānta's logic), the petty officials to the lackeys of the court system. Thus Kamalākānta is offering a defence in a larger sense. Subtly and indirectly, he is less the witness, more the accused, or if he is witness it is to a defamation of something larger, incomparably more significant than the individual and trivial cow. Actually, the scene shifts subtly from the theatre of this everyday courtroom to the general theatre of history where Kamalākānta with the occasional help from Prasanna, is concerned, with his meagre resources, to defend an entire civilization.

Why is the court his favourite enemy? Why do Kamalākānta the essayist and Bankim the novelist always choose the system of justice for their particular ironical attention? Why not something else that is equally undeniably British, or equally clearly unjust? Bankim, it appears, has a reason. Surely, there are other, perhaps greater, theatres of injustice than the British courts. But in a sense, this institution particularly deserves travesty. For the system of justice is itself a parody in the first place, a distinctly un-Naoroji-like idea.[9] Courts carry on the systematic practice of injustice in the name of justice, and the moral and institutional improvement of colonial people. That is why it is the real centre of colonial ideology and is an eminently suitable candidate for the unmasking of its pretensions. Earlier we have heard of Kamalākānta's wish, disclosed in a fit of confidentiality, to the *Bangadarśan* editor, to write a sequel to *Don Quixote*:

> We can write novels quite well; although we thought instead of writing trivial fiction we would write a sequel to *Don Quixote* or *Gil Blas*. Unfortunately,

> we have not yet been able to read either. Would it serve your purpose, Sir, if we write an appendix to Macaulay's essay; for that too is a novel, after all.[10]

Here, at the end of his fictive career, he does exactly that. Or indeed perhaps more: for he does not write a *Don Quixote*, but turns into one and takes on cultural monsters. It could be done only that way, only as parody, as a caricature of the self: in the world of colonialism, it was not in the realm of the possible. In a sense, there is thus an ambivalent climax, both a high point and the low point of Kamalākānta's idiosyncratic narrative: for he realizes here, in a formal symbolism of the theatre what he had earlier merely suggested—a battle between the colonial rationalist system and its accomplices, and the single, vulnerable, unaided native, fortified only by the double conviction of his helplessness and rightness. Appropriate to the double-valued structure of the narrative, this high point is indistinguishable from its point of collapse, for here the text changes, its tone degenerates often into pure farce. It is literal denouement, for Kamalākānta does not appear again. Khosnavis, the babu chorus on his activity, dismisses him with finality: *dekhilām mānuṣṭā khepiā giāche*:[11] I realized the man had become completely, irretrievably mad. Kamalākānta realizes himself by going over a boundary of some kind, beyond the possibility of compromise and translation. From the point of view of collaborative colonial rationality, he was lost beyond redemption.

In this lonely battle Kamalākānta's manner of fighting is interesting. Notice his weapons. There were traditional arguments with which conservative Indians had fashioned a kind of defence, the argument which said that Indian society was different, and western laws, though abstractly fine, did not fit this society.[12] Kamalākānta spurns such weapons. He goes directly into attack, not with 'our' weapons, but 'theirs'. He is out to show—not that western arguments do not apply here though they are sound as arguments, but that they are inconsistent. Rationalism, when questioned from presuppositions different from its own, breaks down under investigation. But to do that, you must be a Kamalākānta, accept loneliness, to have the radical courage to turn his own trial into a trial of the court.

This farcical ending of *Kamalākānta* contains an intriguing questioning of the problem of identity as rationalism conventionally constructs it. On my reading, it also does something far more interesting and fundamental. Kamalākānta was a subtle observer of the complicity of philosophy in language and social practice; and it did not escape him that rationalism claims to use a peculiarly clear language, a language of the literal which holds in unambiguous forms its objective images of the world. Science speaks this clear, literal language, and so does judicial process and law, the field which, like science, claimed ability to reach undistorted truth. Kamalākānta's critique of rationalism must therefore begin at the beginning, in this literal language, in which beyond the judge and the lawyers, beyond even the legal system, his object of ridicule is the linguistic pretence of rationalism. It shows the vacuousness of its concept of identity, a concept or principle taken to be self-evident. His brush with the court here is unusually violent: but that was hardly his fault. For in the first scene, the court, representing western theory, is peremptory and aggressive. It asks Kamalākānta to define his identity, before the actual business begins. Kamalākānta is brought in as a prosecution witness to identify the vital cow, on the basis of Prasanna's well-judged contention (she does not know how well judged it is) that 'this Brahmin never tells lies'.[13] *As a matter of course*, the court asks Kamalākānta to identify himself. Obviously it considered the question of identity a minor routine one, to be inattentively taken for granted. It does not realize the enormous significance of asking a thinking person 'Who are you?' Kamalākānta violently refuses—deconstructing, undermining, ridiculing the concept of identity implicit in the court procedure. He converts this from a mere empirical datum to a philosophical question. It is in that sense, though not aesthetically, a climactic point of interrogation and refusal.

'Who are you?' the lawyer asks Kamalākānta, disastrously underrating his victim. (Literally a victim, '*badhya*', to be destroyed: in his incarnation as the lawyer his legitimate victim is the litigant.)[14] Kamalākānta refuses, and begins to play with this most commonsensical of questions. How do we judge a discourse unless we know who it is that speaks? How can Kamalākānta question this question? How can he see, or pretend that he sees unreasonableness in the most routine opening of a lawsuit? I believe this is the crucial centre of his enterprise. From the critical point of view his approach is unexceptionable, indeed, the only properly critical one. As a lonely critic, in a strange world in which even those he is speaking for are not with him, he realizes that to be a critic is 'to go to the root of the matter' and to go to the root of the matter is to get behind the self-evidence of words and objects. Behind Kamalākānta's apparently casual style there lies thus a deadly serious programme of critical replies. In this we can hardly do better than follow the contest literally.

> Kamalākānta was put into the witness box, and started chuckling. The attendant took him to task: 'Why are you laughing?' Kamalākānta said with folded hands, 'I have not done anybody any harm, why are you boxing me up like this?' The attendant did not see his point, shook his head and said, 'This is not a place for jokes [*tāmāśā*, a clear double entendre for it means both the art of joking and an object of jokes, something that is farcical or comic]. Take the oath.'
>
> A muhuri then began to read out the oath: Say in the direct knowledge of the Almighty . . . .
>
> K. [astonished] What was that you said?
> M. Can't you hear me? With the direct knowledge of the Almighty . . .

K. 'With the direct knowledge of the Almighty! what a presumption!'

The Judge noticed that the witness was causing some trouble and asked, 'What presumption?'

K. Must I say that I had direct knowledge of the Almighty?

J. What is wrong in that? That is the form of the oath, after all.

K. You are a great judge indeed, sir. It could perhaps be excused if I said some minor lies in course of my deposition, but should we begin by such a massive untruth?

J. What do you find untrue in this?

Kamalākānta said to himself 'you would not have come this far if you could guess'. Aloud he remarked 'Your honour, I always had the impression that God was not an object of direct knowledge. I have never yet seen him clearly, directly, but that could be because of poor sight'.[15]

The judge, to make matters brief, said, 'give him a simple affirmation'. Kamalākānta is now asked to say 'I swear', which he refuses on the Austinian ground that he could hardly say I swear without knowing what he would have to swear to. It is, in other words, a performative; and it was rather like getting someone to sign on a blank paper, which, he says archly, is a common practice under the colonial administration. 'I knew this was done outside court all the time, but did not realize this was court procedure as well'.[16] Beginning his interrogation the lawyer asks what is his name, and what is his father's name, leading Kamalākānta to enquire if even depositions must have genealogies. 'Which caste are you?' which, reading it literally, he answers by asking how could he, an individual, constitute an entire caste.

Which category? — Of the Hindu category.

Which *Varna*? — Very dark [because *Varna* means both colour and caste].

Have you got a caste? — Who can take it away?

[Asked his age, Kamalākānta says] 'fifty-one years, two months, thirteen days, four hours, five minutes . . . '

What nonsense. Who wants your hours and minutes? — Why? Did you not make me swear just now that I would not withhold any information?

Do what you will. I can't handle you.

Where do you live? — Nowhere.

I mean where is your home? — I do not have a room to call my own, let alone a house.

Where do you stay, then? — Wherever I can.

Don't you have an address of some sort?	I did when Nasībābu was alive, no more.
Where are you now?	Why? In this courtroom!
Where were you last night?	In a shop.

The Judge butted in and said, 'Let us stop this debate, let me write down "lives nowhere".'

What is your profession?	Am I a lawyer or a prostitute that I must have a profession?
I mean, how do you manage to eat?	Usually, mixing rice with *dāl*, using my right hand to lift it, and direct it into my mouth and down the throat.
How is that rice and dāl provided for?	If God provides them, not else.
Do you earn anything?	Not a penny.
Do you steal, then?	If I did, I would have sought your assistance sooner, and you would have got your share of my earnings.[17]

The lawyer fails. He gives up and tells the court he would not be able to question this witness. Symbolically, intransigent helplessness wins over power. For all the majesty of its power, the rationalist system of the court fails in its main dramatic purpose; it cannot bring Kamalākānta to submit to its discourse, to speak it, to answer its questions the way it wants. Rather, what happens is just the opposite, the questioning of the victim, Kamalākānta, breaks up the court's discourse, destroys its two crucial pretences of objectivity and consistency. The discourse of colonialism, and its supporting discourse of European Enlightenment, are shown to be hollow and eventually vulnerable; its majesty holds only as long as its power exists, only as long as its interlocutors do not question its assumptions. Colonialism admits of no rational defence. Officials of the court make an indistinct crowd, formed partly of a judge, the lawyers, the transcribers, the ushers—all of whom are seen partially as in a painting of a crowd; their thinking and speaking are hazy, and they are entitled to only the haziest of existences. Against them, figurally, Kamalākānta stands out clear and undeniable: only he who takes responsibility for his words, who goes to the primal quality of their meanings deserves to be noted at all. The qualification for being a serious critic is set out sharply—a critic has to give to his vocation his life, not his spare evenings. Criticism is not a way of saying things, but a way of being in the world.

We realize immediately that this is a particularly powerful instrument of caricature, for it makes fun of the fundamental claims of rationalist discourse. Caricature of this sort does not merely show gaps or infelicities in rationalist doctrine from the outside; it does something more lethal.

It is an inextricable part of the rationalist project to clarify language, and through language its relationship with the world. It sees itself as a force that cleanses language, destroys superstition, and is entirely internally consistent. To show that rationalism itself stealthily commits what it condemns in others, to demonstrate that it is not adequate to its own declarations, is a particularly damaging criticism. Kamalākānta's criticism of the enlightened court is mostly of this form. The theatre and conventions of colonial judicial practices were meant to induce belief that the process going on was unbiased, an entirely rational judgement about a case. These conventions are derived from prevalent philosophical doctrines, suitably dramatized and abbreviated into a popular language. The judicial procedure draws its claim to immunity and eminence from philosophical theories of truth. Kamalākānta's early remarks reveal that this language and the philosophy that confers respectability on it, are full of unexamined propositions. Thus a theory which prides itself on its freedom from superstition begins with superstitious beliefs; a theory which ridicules the practices of others has ridiculous practices of its own. A system which declares its objectivity is biased in its own initial moves.

Examples are carefully constructed so that utterances have a routine meaning within the court language,[18] but outside the language of the court they still retain enough of an identifiable meaning, but a meaning that is offensive, laughable or goes wild. The whole scene illustrates how, although they speak ceremoniously to each other, and are structured into the same face to face situation, they are not able to effectively communicate. Kamalākānta's perverse consistency in misunderstanding whatever the lawyer says underscores this lack of communication. A word like *sambandha* (relation),[19] with a purely neutral connotation in law, means a simple acquaintance. In ordinary discourse of Bengali social life it could mean a marital, and in special innuendo, extramarital relation, the sense in which Kamalākānta takes the word, to create an excuse for taking offence. A lighter game is played on the word *ṣāmlā* [20] indicating indistinctly a black cow or the lawyer's gown. This comedy of errors works on an interesting principle. In hermeneutic theory, interpretative superiority is ascribed to that view which understands the view of the other, without its obverse being true. The situation in *Kamalākānta* is exactly similar: those who speak the indigenous discourse must, because of their subjection, understand the English rationalist language game. But the British, or the babus who inhabit the discourse of the colonizer and are used to its privileges, need not do so. Therefore, while what the lawyer or judge says is clear to Kamalākānta, although he creates impenetrable incomprehension precisely because he understands, this is not true of the other side. By a literary route, Bankim comes to emphasize the idea that the subaltern, the dominated, must have a clearer view of the world, an idea that could almost be taken from Hegel's *Phenomenology*.

At the end of the episode, Kamalākānta is back to the question of colonial conquest in a reference that is at the same time explicit and travestied; the minor affair of the cattle theft is regarded as a small but typical example of the historical principle involved which makes aggrandisement and justice indistinguishable.

> Long ago a brahmin said to King Śyenajita, between the owner and the thief of a cow whoever drinks its milk has a stronger claim. For others to express any attachment to it is importunate. This is the law of the Hindus as expounded by Bhīṣmadeva Thākur; this is current European international law. If you wish to be regarded as civilised and advanced, you must snatch things away from others. Whether the word *go* means the earth or the cow, she is destined in either case for the enjoyment of thieves. All thieves from Alexander to Ranajit Singh illustrate this rule. If the right of conquest can be a right, why not the right of theft? Therefore, O Prasanna, act according to law. Follow the lessons of the history of diplomacy. Let the thief take the cow.[21]

If his answer to the court is not read as a series of no-sayings, like stating I would not say where I live, or more radically, I do not stay anywhere, or connecting all such non-answers and thus undermining the court's conception of rationality, one perhaps finds something more interesting in the meaning of his madness. His intransigence could refer subtly to the condition of being *aniketa*,[22] someone who refuses a parochial home in order to live in the world. It is then a positive philosophical statement, which the court, because it is unacquainted with Indian thought misunderstands in a way both comic and infuriating. The ideal of the aniketa even though translated to the hākim, would only elicit the astounding mistranslation *nibās nāi*.[23] This is typical: it is both a literal transliteration of the word and a total incomprehension of the concept.

It is inadequate to see him as merely non-stating, playing a farcical game of wrong answers; in fact, he is playing a game of a far more serious and deeper kind with the court. There is an underlying system of description of the self, but so coded that it might be missed by its two audiences, the internal audience of the court, but also by people like us, to the extent we are like the people inside the court. The appellation of madness by the dominant discourse is a way of creating a privacy for oneself. Within his game, it makes perfect sense to say and not say exactly what he does: that he does not have a profession, no residence, no income, none of the fixed bourgeois references which are supposed to constitute a man's humanity. They do so contingently, only in this social arrangement, and do not have any transcendental title to truth. Kamalākānta asserts, in effect, you may know a person's income, profession, address, the entire list of characteristics which, on this theory, constitute his personality, yet not know him. By this, he is creating space within language to say both things, an enormous terrain of defence and interrogation.

Through this indecipherable complexity, a language in which we are never able to reach the bottom, he is perpetually absconding within these words.

The Babu in Synchrony and Diachrony

Descriptions of a society could be of two kinds, which could be known/recognized by two languages. One was the description of the society in an insider's language: which simply produced, enlarged and extended its *ideology*, its self-justificatory consciousness. Another description could be through the language of an outsider; not a natural outsider, who simply stands outside a society as an alien. The right to speak the truth belongs to a person who has worked his way to the outside; he has *become* an outsider, to him belongs the right to see its truth. In *Kamalākānta*, there are, apart from the unclassifiable commentator himself, who does not speak the language of any particular group, some other characters. Between them, they describe the social geography of colonial Bengali society. There is Prasanna, the symbol of the *feminine*, with the complex addition that she also represents the *popular*. There are the indistinct collective portraits of the court men. These are of course paradigmatic babus, displaying their quintessential features. The characteristic that distinguishes the babu in Bankim's time and ours, is crucially the acquired unfamiliarity with one's own culture, its language and its discourses under the misleading belief that one has mastered the superior ones of the West. The men of the court are insignificant superstitious rationalists (rationalists without knowing why it is good to be a rationalist), which is their collective structural property, quite apart from being, as individuals, disingenuous rascals. But beyond the babu one gets an occasional glimpse of the third element of Bengali society, the typical British colonial administrator. It is between these three figures that the entire social geography of colonial Bengal is laid out. I shall treat these types in succession: first the babu, the real object of description and travesty in the text. Second, we shall open up the condensation in the character of Prasanna, and treat her two different characteristics distinctly, the symbolism of the woman, and the role of the popular characters in Bankim's humour. Although the popular do not erupt in Bankim's descriptive world too often, when they do, they appear as representatives of a laughing people.

The great object of attention of the *Kamalākānta* text is the babu. He is the constant object of Bankim's humour, in all its various moods, from the vicious, to the gentle to the forgiving. And, not surprisingly, the babu is not a new theme with Bankim, won for the dazzling show of wit in *Kamalākānta* : indeed, he is in this somewhat wry sense, his first love. Two of his earliest pieces of humorous writing immediately and unerringly identified this abiding object of his sarcasm, the babu, the collective type with whom the author has such a fertile relation of contradiction. He is undeniably a part, and at the same time, he could not accept he was a part, leading to the foundation of the great tradition of Bengali literary self-irony. *Ingrājstotra* (hymn to the Englishman),[24] helpfully subtitled 'translated from the *Mahābhārata*', the all-seeing chronicle of India comprising its past and future, at once established both the form and the content of this humour. The *stotra* would undergo unending experiment at Bankim's hands, running the whole gamut of sentiments from the ridiculous to the sentimentally uplifting.[25] He was indeed to reshape this most fundamental form of invocation in the Hindu tradition to startlingly novel purposes. To be a stotra a composition must however conform to some purely formal properties of style. Incomparability of the deity is conveyed by the mannerism of descriptive excess.[26] Stotras usually have a circular, repetitive style, coming back after each cycle of excessive praise to the signature phrase. In Bankim's early travestied uses of the stotra form there is a certain debasing of its style which can come only with an easy and shrewd familiarity with its formal properties, just as a successful cartoonist would create laughter by exaggerating credible features of a face. Early parodies like the *Ingrājstotra* are therefore convex satires which pour sarcasm on the Englishman, the object of the incantation, the babu, the reciter whose discourse it encapsulates, but also, subtly, the doctrine of excess in the stotra form. Formally, it immediately affiliates itself to Bankim's favourite, the *alaṃkāra* of vyājastuti; and its content is a double description; a description of the Englishman in terms which vividly describe the describer, a self-characterization of ascending/intensifying servility.

> O one who can divine what is going on inside our minds! whatever I do is meant to win your heart. [Though the Bengali verb '*bhulāibār janyya*' is more double-edged, and means equally, to deceive you; so the correct rendering of meaning would be: to win your heart by deception.] I give to charity because you call me an altruist; I study so that you may call me learned . . .
>
> If you so wish [or, because you wish it] I shall establish dispensaries, for your approbation, I shall set up schools, according to your demands I shall give subscriptions. I shall do whatever you consider proper. I shall wear boots and pantaloons, put spectacles on my nose, eat with knife and fork, dine at a table, please have pleasure on me.
>
> I shall renounce my mother tongue to speak your language, abjure my ancestral religion and adopt the Brahmo faith, instead of babu use Mr as prefix to my name, be pleased with me.
>
> I have given up my meals of rice and now eat bread; I do not feel properly

> fed until I have some forbidden meat: I make it a point to have chicken for snacks, therefore, O Englishman, please keep me at your feet.
>
> Please grant me wealth, honour, fame, fulfil all my desires. Appoint me to high office, or a raja, or a maharaja, or a rāibāhādur, or a member of the Council; if you cannot grant all that, invite me, at least, to your at homes and dinners, nominate me a member of high committees or the senate, or make me a justice or an honorary magistrate . . . Please attend to my speeches, please read my essays, give approbation—then I would not pay attention to the denunciation of the whole of Hindu society.[27]

Clearly, there are two levels of meaning in this false hymn. At the first level, there is a caricature of both the babu and the Englishmen who confer honour on him. Characteristically, Bankim goes immediately to the heart of the matter, cutting through the pretences. Only in appearance is colonial society a field of career open to talent, an achieving society; at bottom, the colonial administrative system does not encourage meritocracy. The Englishman can give anything; it is the arbitrariness of his conferments which is emphasized, and which makes the babu's supplicatory self-abasement a proper complement to it. High honour under colonialism in Bengal is not recognition for desert, service or ability but for competitive self-abasement. Clearly, colonialism confers on the Englishman a mystical power of nominalism; he can name anything into anything; and the main point is to be so named by the right authority. The English can rename all social and moral descriptions: and most tellingly of all, the babu's adoption of reform and rationalism is revealed for what it is. He is a rationalist out of opportunism, and entirely unclear about how a rationalist argument is grounded: he would do all the right things—accept modernity, break tradition, adopt altruism, not because he can ground it himself in reason, but because the English would consider it praiseworthy. The babu's acceptance of western rationalism is fundamentally tainted by this heteronomy. Thus two types of acts can be behavouristically indistinguishable: but whether it is an act of altruism or servility can be found only by looking into its rational authenticity. The upside down character of the colonial world is etched in briefly and powerfully through this supplicatory refrain: *I will do everything you ask for*, turning the right acts into wrong ones. This is why colonial society is such a proper field for a sarcastic demystification. Even seemingly high-minded action must be probed by this sarcastic mistrust, until they reveal their true motives. It is the unapparent, indistinct intention which can tell an act of kindness from an act of imitative servility, or simply verbal posing from genuine intellectual curiosity.

This was an early piece. Compared to the subtlety and power of his later caricatures of the babu discourse, this piece appears unrefined. Its significance lies in the fact that it sets a pattern, a fundamental descriptive structure, and it is interesting how little the main point of the structure of babuness would change in his mind. In sheer literary quality it is not of the standard of the first, an elegant essay on the learned tiger of southern Bengal who delivers a public lecture on his anthropological observations of human society when he had strayed out of the forest and spent an instructive period in the Calcutta zoo.[28] But later, in *Lok Rahasya*, he writes a piece of such sustained accomplished irony that I doubt if even Bankim ever matched its vitriolic excellence.[29]

Like the hymn, this too is purportedly taken from the *Mahābhārata*, using it and travestying it at the same time, turning its claims to allseeingness against itself. Vaiśampāyana, the sage who recited the epic at the court of Janmejaya is caught in the early parts of his epic performance; and the king, a man of great historical curiosity, requests the sage to recite to him the qualities (*guṇa*)[30] of those who would adorn the earth in the nineteenth century and would be known as babus. This piece is particularly resistant to translation because of its heavy reliance on the form of *vyājastuti* manipulating the ambivalent associations of phrases and straight double meanings of terms in Sanskrit. Though Bankim used such pieces of modern *vyājastuti* later, few indeed would come anywhere near this unparalleled intensity of humour and ambivalent denunciation.

The author and reciter of the *Mahābhārata* enjoy the reputation of being *sarvadarṣī*, all-seeing. And this is not in vain; for he compresses the historical characteristics of the babu into an unsurpassable character construction. It founds something like a distinct genre of praise-abuse of the self of the Bengali middle class bhadralok.[31] Only the rise of more serious nationalist politics puts an end to this humorous literary tradition.

An approximate idea of Vaiśampāyana's characterization of the babu could be found in some of the passages, though a translation would miss the insistence of the series of adjectives in Bankim's use:

> Babus are invincible in speech, proficient in foreign language and hate their own; indeed, there would appear some babus of such amazing intellect that they would be altogether incapable of conversing in their mother tongue . . . The babus are those who would save without purpose, earn in order to save, study in order to earn and steal question papers to do well at examinations. Indeed, the word babu would be many splendoured in its meanings; those who would rule India in the *Kali* age and be known as Englishmen would understand by the word a common clerk or superintendent of provisions; to the poor it would mean those wealthier than themselves, to servants the master. I am however celebrating the qualities of some people whose only purpose in life would be to spend a fittingly babu existence. If anyone takes this in any other meaning, his hearing of this *Mahābhārata* would be fruitless;[32] in his later existence he will be born as a cow and constitute a part of the babu's dinner . . . Anyone who is devoid of any poetic sense, with an execrable musical sensibility, whose only knowledge is confined to textbooks read in childhood, and who regards himself as omniscient is a babu . . . Like Viṣṇu they would

> always lie on an eternal bed. Like Viṣṇu, they too would have ten incarnations; clerk, teacher, Brahmo, broker, doctor, lawyer, judge, landlord, newspaper editor and idler. Like Viṣṇu, in every incarnation, they would destroy fearsome demons. In his incarnation as a clerk he would destroy his orderly, as a teacher he will destroy the student, as station master the ticketless traveller, as Brahmo the small priest, as broker the English merchant, as doctor his patient, as lawyer his client, as judge the litigant, as landlord his tenants, as editor the ordinary gentlemen, as idler the fish in the pond . . . Anyone who has only one word inside his mind, which becomes ten when he speaks, hundred when he writes and thousands when he quarrels is a babu. He whose strength is one time in his hands, ten times in his mouth, a hundred times in his back and absent at the time of action is a babu. He whose household deity is the Englishman, preceptor is the Brahmo preacher, scriptures are newspapers, and place of pilgrimage is the national theatre is a babu. One who gives himself out as a Christian to missionaries, as a Brahmo to Keśabcandra,[33] as a Hindu to his father and an atheist to the Brahmin beggar is a babu. One who drinks water at home, alcohol at his friend's, receives abuse at the prostitute's house, and kicks at his employer's is a babu. . . . He who hates oil while taking his bath, his own fingers when eating, and his mother tongue in conversation is indeed a babu . . .
>
> O King, the people whose virtues I have recited to you would come to believe that by chewing *pān*, being prone on the pillow, having bilingual conversation, and smoking tobacco they will regenerate their country.[34]

Janmejaya, apparently an astute observer of men, had a fairly clear idea of what sort of beings the babus would be, and requested the sage, 'O sage, let the babus be victorious, please turn now to some other subject'.[35]

I have on several occasions mentioned a stylistic feature of Bankim's prose—his deliberate use of Sanskritic tropes, and his general nearness in language use to the classical literary tradition. Nowhere does this come out so clearly as in this piece. His humour is inextricably related to the pleasurable discovery of the idiosyncracies of language, the capricious associations of its idioms, nouns, verbs. Entirely dissimilar things sound linguistically, or rather phonetically similar.[36] This is stuff out of which poets would conventionally create the end-rhymes of their verse. Classical poetics realizes shrewdly that there is a place for such play of phonetic similarity within prose styles too, creating startling sound effects and pleasantly capricious affinities. Bankim literally revels in these startling resemblances. He loves to explore, as the musician does in the case of notes, the internal continent of the language as a world of constantly changing, constantly surprising resemblances—of meanings, sounds, of rhythmic patterns. Bankim belongs to a period of literary Bengali heavily modelled on classical Sanskrit literary norms which is soon left behind. Puns become distinctly less reputable in a few years within serious prose. Humour, by the time we come to Tagore, is nearly all semantic rather than verbal.[37] The locus of the ironical is shifted from the sounds to the meanings, from the sensuous rustle of the language to the ideas connoted.

If we analyse all his characterizations, for Bankim, the babu's primary feature, which is instantiated in varying ways, is inauthenticity. The babu is a sudden and non-autochthonous arrival on the social scene. He was not implicit in any earlier development in Indian or Bengali society; he derives his sudden appearance and immediate eminence to colonial conquest. So the babu is not the result of his own history, in a sense; he is simply a contemptible corollary of British conquest of Indian society. Inauthenticity is reflected in different aspects of the babu's existence, and two fields are taken up for special attention; the babu's pretensions of power based on a rationalist conception of the world, and his cultural existence, his claim to be cultivated intellectually.

The rationality of the babu which allows him to declare his flimsy rebellion against traditional Hindu society is derived from his conversion to a kind of discourse, made possible by the new, western education. But there are two senses in which this rebellion is hollow. Intellectual rebellions too have their political aspects; the babus did not work through the logic of a rebellion against Hindu society. They were the products of the great externality of Indian history. Their rise did not lie in anything internal to the process of Indian events. Affiliation to British power made them both vulnerable and irresistible. They could change everything they wanted in traditional Indian society, not by their own power but of those for whom they had chosen to become cultural brokers. This peculiarly vulnerable, shaky omnipotence of the babu alternately produces mirth and indignation in Bankim's literature.

Bankim lived at a time when the ambitions of the babu were primarily cultural and reformist; but he could see that it was already becoming political at the edges, something Bankim could only laugh at. He found even pretensions of reformism (of the Brahmo variety) ridiculous; no wonder he would consider the idea of more adult political ambitions of this dependent class a sordid joke. Politics, he implies, had two great sources; it could arise out of real power, or out of a genuine, large, historic resentment. Babus were mere heteronomous agents of others' ideas and others' acts. Consequently, they lacked either the power to really restructure and rearrange their own society, for that kind of renewal too needs courage, the sort of courage that the politics of the Enlightenment demonstrated in European history; or the responsibility of being the focus of a society's historical sufferings. Both these could arise out of being part of the society's logic of political or economic reproduction, what could be called, in a different way, greater authenticity. But this class lacked both these distinctions. As a result, they were vulnerable from two sides; they did not know where they stood with the British, and their position could change drastically with the merest change in administrative policy of the colonial power. They did not know where they stood in terms of their relation with the indigenous society. Not accepted as part of the British colonial power (culturally), rejected as part of the society they had voluntarily disowned, they represent the pure form of heteronomy. For such a heteronomous group, so obtuse as not to see the racial basis of colonialism, not quite Indians, politics was a ludicrous proposition. Naturally, it is hardly surprising that with true suicidal instinct, the editor of *Bangadarśan* requests Kamalākānta for an article on a political subject.[38]

The editor asks for an article on politics because apparently he would use this as part of his political strategy; as a fearless Bengali journalist, he always invites others to fight for the principles he holds most dear to his heart. 'Under the new laws, there would be a shortfall of supply of political writing from other sources',[39] and Kamalākānta is asked to provide some from his uncensorable pen. Kamalākānta snorts with indignation at the suggestion, and denies all connection with the world of politics. He does not, he asserts, belong to the directory of politicians. His rambling letter then bifurcates two definitions of politics, one that sees it restrictively (the babu way; i.e. politics as what happens in meetings, speeches and newspaper articles), and the other which sees it as universal, something that is inescapable and goes on everywhere (the non-babu way).

> Am I a landlord or a sycophant or a fraud or a beggar or an editor that I should start writing on politics? You have I assume read my journal. Where did I display such imbecility that you ask me to write on politics, of all subjects? True, I did occasionally flatter you to secure my supply of opium; still, I have not degenerated to such a self-interested sycophant that I should write on politics. Fie on your editorship, fie on your gifts of opium, you have not realised that Kamalākānta Śarmā is a highminded poet, not a mean-spirited dealer in politics (politician).[40]

But Kamalākānta is a prisoner of his second sight, which cannot stop seeing; it reveals the truth (or the logic) of the world, even when he turns away from it. Exactly this happens to him when he turns away from the irritating prospects of political journalism. Turning away from the subject, he looks at what he initially regards as an ordinary everyday village scene—a courtyard of the house of a poor oil-presser (*kolu*), his cattle feeding from their respective pots of hay, his son noisily prosecuting his lunch, his wife generally in charge of the peace of the household. This ordinary sight changes, and he makes a startling discovery. In his reverie he continues to think absently of politics and is reminded of a famous ditty from a folk play (*jātra*), a travesty of a classical *sloka*:[41]

> the dumb wishes to break into words
> the lame wishes to run about
> you likewise wish that learning should occur to you
> remarkable wishes! etc.[42]

Interestingly, we find something that we shall come across repeatedly. The voice of truth is a mocking, ironical voice, and the privilege of speaking it is accorded to folk. We can half guess the historical reason for this. Through the unfortunate history of his society, the high culture has become decadent, nearly mute. The new culture is a culture of the inauthenticity of colonialism, it simply cannot have the right to speak what is true. That can belong only to the folk, the voice, the mockery, the ironies of its people. Besides, in the original sloka '*mūkaṇ karoti bācālaṃ/ paṅgurlaṅghayate giriṃ*', it is a mark of the omnipotence of God that he can make the mute garrulous and the lame cross a mountain; and obviously, through the slide of associations, for the babu this can be done only by the Englishman who acquires through colonial culture a power, equal to God, of making the impossible happen, Thus, he continues:

> We similarly wish to do politics, politics every week, everyday, every hour. But exactly like the wish of the dumb for oratorial excellence, the desire of the lame to travel fast, of the blind to see paintings, like the wish of the Hindu widow for the affection of her husband, like my wish for the love of a devoted wife, it is ridiculous, never to be realised. Brother dealers in politics, I, Kamalākānta Cakravarty, am advising you for your own good, even the impossible can happen, but a nation that allowed itself to be conquered by seventeen cavalrymen, can never achieve politics. 'In the name of the Lord, give me some alms'—this is all the politics they can ever have.[43]

This passage and its tone is typical of Bankim: its mixture of sarcasm and a certain melancholy. It announces the impossibility of politics, but the dialectic of his thought immediately sees the opposite. After this intense discharge, he keeps looking at the scene in the oil-presser's house, depressed. Under the influence of opium, he saw something else metaphorically inscribed on the scene: a dog sidled up to the boy and started wagging its tail with demonstrative loyalty, and the boy duly rewarded him with a bone he had picked clean. Kamalākānta realized now that this dog was a politician; every move of his was political in character. The dog found that the boy took no further notice of him; so he moved up closer, started wagging his tail and putting out his tongue began to gasp. 'Looking at his emaciated body, empty belly, beseeching eyes and quick gasping, the boy felt pity; taking a piece of fishbone he sucked it clean and threw it to him; the dog's political agitation thus met with success. The dog, beside himself with joy, started munching, licking, swallowing and digesting it. His eyes closed with pleasure'.[44] The dog grew bolder, and after a time started whining and moving forward slowly. The boy then ran through the fish and threw a gob of rice at him. The boy's mother came out, and finding a strange dog sharing her son's lunch threw a piece of stone. 'The politician, thus hurt, gathered his tail, and to the loud accompaniment of various classical *rāgas* made himself scarce with great speed'.[45] At the other end of the courtyard, a more stout ox than the oil-presser's had come in and started eating his meal. The oil-presser's wife took a stick and chased the intruding ox. Without moving, the ox lowered its horns, and 'intimated to her the strong likelihood of his horns entering her heart' if she persisted in her pursuit of justice, upon which the lady retreated and left him to conclude his unfinished if not wholly legitimate meal. 'I thought this is politics too. Actually, I was shown politics of two different kinds; one a politics fit for dogs, the other for oxen. Bismarck and Gorchakoff are politicians of the ox variety, and from Cardinal Wolsey to our own Mucirām Rāy Rāy Bāhādur are politicians of the dog version'.[46]

For the colonial mind, denied the liberty of speaking about politics, it spreads across the world in a gigantic and ineradicable metaphor. Kamalākānta does not speak about politics, but sees politics written across the universe, and is endowed with the vision to decipher the hieroglyph of power in everyday life.

Political ideas had defensible and rational definitions within the structure of western rationalist discourse. But the babu's adherence to them was purely and peculiarly external. He acceded to them not because they were rationally justifiable; indeed, he would have been found wanting on the first principle of rational justifiability, which can be established only through an initial admission of sceptical doubt. The babu did not know how to defend the rationalist corpus rationally, on his own; he took the entire apparatus of rationalism on trust, because the European believed in them. He had thus, an entirely irrational belief in rationalist theory. Not surprisingly, this travestied rationalism gets a hilarious critique from Kamalākānta in one of the essays.

> Learning: it is difficult to determine what learning really is. To some people, it is the ability to read and write. Others believe that to be learned it is not necessary to read and write a great deal, all that one must learn is to write books and publish in newspapers. Some others differ, arguing that if someone does not know how to read or write how could he possibly contribute to newspapers? I consider such debate insignificant. Young crocodiles jump into the water the moment they come out of their shells; they do not have to learn swimming. Similarly, to Bengalis learning comes naturally, they do not need to learn to read and write in order to be learned.
>
> Intellect: the amazing ability by means of which we can regard iron as cotton wool, and cotton wool, iron. Like the hoarded riches of the miser we can always see it, but it is invisible to others. Of all created things, this must be the most plentiful in the world, for I have never heard anyone say that he has less of it than others.
>
> Labour: to partake of properly heated delicacies, followed by sleep, followed in turn by taking the air, smoking the pipe, conversation with one's wife—strenuous activities of this kind are generally called labour.

> Power/strength: Long-winded sentences, flushed face, great verbal noise, a constant shower of Hindi and English words accompanied by spit, gesticulating kicks and blows from a reasonable distance, and untimely exit at the slightest move by the adversary—such things are generically called strength.[47]

The sheer persistence of two themes in Bankim's rejection of the babu is remarkable. First, the inauthenticity of the entire intricate structure of the babu's thought, its rootlessness in his own experience, something intensely, obscenely, infuriatingly false, but for which we have not as yet been able to find any appropriate name except the very abstract one of heteronomy. Theoretical constructions came on from the outside permanently marked by this flimsiness, this insignificance of exteriority, of not being a result of reflection of a deep, wide historical experience, but a dishonourable result of cultural imitativeness. The same idea, an identical sentence, spoken by a European author meant one thing, when spoken or rather uttered by an Indian babu something culturally quite different. The two sentences may be exactly identical; yet one is a caricature of the other. The second remarkable feature is the constitution of the babu almost entirely by his speech. The babu is not a creature of practice, but of words. After devaluing his stock of words, the destruction of his form of humanity is complete, for he never left, in any case, any practical residue.

I have said earlier that I would not treat *Kamalākānta* as a text in the limited sense of the single book. Text means a weave, and things could be woven into visions in endless different ways. I have also tried to show earlier how difficult the author found the conclusion of *Kamalākānta*; it had several endings after which it began again. Since Kamalākānta is Bankim's oblique vision, a crooked way of seeing to which the devious world reveals its truth this vision irrupts everywhere in Bankim's work; for he is his other authorial self, as inseparable from him as a shadow.

Kamalākānta represents in that sense less a book, than a mood, a way of seeing and relating to reality. And anything that is a product of this way of seeing I assimilate unrepentantly into the larger text of *Kamalākānta*. By this I assimilate two other texts into its typical weave; the short skits of *Lok Rahasya* and *Mucirām Guḍer Jīvancarit*, as they are visions of colonial life produced by the same eye, the self of which he had declared: 'I am an opium-eater; my hearing, understanding and writing are all out of tune'. (*Āmi āphingkhor, besuro śuni, besuro bujhi, besuro likhi.*)[48]

However, one must note the deeper, cruder, angrier tone of the travesty in *Mucirām* which separates it from the much finer texture of *Kamalākānta*: and this makes it, in my judgement, inferior to that more complex, less organized text. The satire in Mucirām is radicalized in one sense. It is surely a more explicit, intense and brutal parody of the babu: however in becoming harsher it also becomes less generally applicable. Obviously, the hesitations, cowardices, invisible betrayals in *Kamalākānta* are small enough to be seen as the secret crimes of all babus. The overt larcenies of *Mucirām* are less widely applicable; and thus, by radicalizing the indictment, it acquits a majority of the defendants. All babus were not frauds of the Mucirām variety, and therefore if Mucirām is held up as a portrayal of the ordinary babu rather than the broken images of *Kamalākānta,* it becomes possible for the large majority of babu society to disclaim all serious kinship with him, and to feel exempted from the point of this critique. Mucirām's story is less effective as collective self-irony. It would fail to make the average babu worry that what was being painted was uncomfortably close to his own portrait, what he secretly knew himself to be.

The Babu's Others: Women

The spectacle of *Kamalākānta* has three major protagonists; the babu's character is constructed dialogically, through a series of descriptive and discursive contrasts with his *others.* From the artistic point of view, the figure is not complete without setting out his relations to his *others.*[49] We must therefore devote some attention to the question: who are the others of the babu? What are the principles of their construction in Bankimchandra?

These others of the babu do not have a merely conceptual existence. Bankim's art mirrors the historical situation of his period of transition. These others of the babu are not brought in and created for the purely philosophic purpose of illustrating that there were other possibilities of existence, other ways of living one's life. They were his historical neighbours too; actual social types who surrounded and interacted with the babu, and in real life he had to contend with them. They are permanent, ineradicable presences in his world. The first others, less important for our purposes, is of course the dominating figure of the British, presented in a double caricature. Sometimes, the British are presented as civilians, administrators who are well-meaning, but whose efforts at reforming from the outside a society which they merely dominate but do not understand end invariably in farcical results. Often however, the British are also presented in their other incarnation, the anthropologist and indological scholar who knows this society better than its inhabitants themselves and is giving it a historical chronicle for the first time.[50] Let us however turn to the other others of the babu. They are, first, his *women*, mostly his wife, the most intimate of his enemies, who represents in Bankim's world a different cultural principle, a different discourse, quite different standards of rationality and intelligibility. Besides, against the babu in the space of the transitional society of Bankim's Bengal stands the figure of the *popular masses*, the average lower order individual, still bearing in her the principles of the cultural world before the advent of colonialism and the babu. Since both of these types, women and popular characters, represent negations of the babu, it is not surprising that often they merge into individuals who display both these traits. The best example of this is of course the figure of Prasanna, the foil to Kamalākānta in the text.

Prasanna is not the only representative of this principle in Bankim's work. In his novels, especially those which I shall call novels of praxis, we shall come across figures whose fictional attributes are different, but who share this cultural principle of non-babuness, and its connection to praxis. Notice the sharply non-babu, i.e., indigenous education given to Śānti[51] and Devī Caudhurānī,[52] and that training, unlike the crippling, immobilizing consequences of babu education, assists them in political praxis. Thus, quite generally, the woman represents a different principle of life—more spontaneous, popular, indigenous, very close to the resources of the folk, against the cultural inauthenticity of the babu. In several novels, women are the enemy of babuness within the home. Consider Indirā and Suhāsinī in the house of the babu lawyer;[53] or differently, Labangalatā.[54] This is brought out far more sharply in the sketchily drawn women of his satire, particularly *Lok Rahasya*. Think of the sceptical wife of the conversation on the exalted occasion of the English new year, the wonderful play on the comicality of imitation (notice the structural parallel, which is the essence of every comical or satirical homology, between the stated imitation of the son in saying exactly what his father has said,[55] and the unstated structure of the babu saying exactly what his superior says). Speeches of superiors change their meanings subtly, inescapably, when they are used by subalterns. At best they turn into lisping imitativeness, or grotesque insults as it happens in the case of the son.[56] Or consider the dialogue, whose parallels must have been going on within every Bengali home of Bankim's time everyday, about the native language.[57] Indeed, few other writers of his generation show Bankim's sense of a moving temporality, the changing present, already giving way to something that is a differently structured social culture. Few others have this fine sense that the least noted elements of social life, its language, its clothes, roles, stereotypes, possibilities of lifestyles are decisively changing. Indeed, I have argued earlier, the babu (or the heteronomous cultural principle) gradually conquers the social world; and the first token of this is his success in the taming of his women. Wives increasingly come to be cultural counterparts of their husbands; they exemplify the same cultural principle, the same norm of cultivation, only in a different, feebler gender. They feel the same sorrows and joys and express them in the same language. Indeed, the world of language becomes far more standardized, uniform, with even servants in Tagore's fiction and afterwards speaking an unrealistic and disappointingly chaste form of middle-class prose. The authors require, of course, some semiotic markers of difference between the language spoken by the master-babu and his servant; but this is shown as a *lack*, as a failure of the servant to speak the chaste Bengali of the babu. This is an entirely different situation from the one in which his language is not marked by a lack, but precisely by a fulness, by a great spontaneous competence. The women in Bankim do not speak the language badly; they speak *another* language exceedingly well. Its expletives do not arise from a fumble at not finding the right word; rather they arise out of a vocabulary that exactly matches their experience, a point Bankim makes quite explicitly in the essay 'Old Women and the New'.[58] Bengali women in fiction in the nineteenth century undergo a transformation of their character parallel to the mythic Rādhā's. From aggressive, self-assured, rural individuals (who are not weighed down by feelings of inferiority to their male neighbours), they turn like their husbands into delicate urban parasites. Once that transformation is complete, they are unable to become the focus or the fictive symbols of a critique within the household, often extending to their rationalist spouses when they are prone to excesses of a subaltern rationalism, the criticism of household weapons.[59]

Nowhere in Bankim's novels is there room for a woman of the new type, who increasingly represents the new cult of feminity.[60] Rationalist influence, ironically, turned women from subjects, albeit unrefined and abusive, into objects of admiration, ciphers of a wailing vulnerability. Bankim's heroines are mostly powerful, aggressive, masterful women who are able to take their own decisions, and work their way against a usually unreasoning and oppressive society. In actual Bengali society however, he was conscious, there was a fast transition to the Victorian feminine. The battle that he helped women win in his fiction they had already lost in society.

Bankim's consciousness shows two aspects of the historicity of such change, interestingly combined. There is a sense of the short-term inevitability of this change, a clear-sighted apprehension that such bad things are going to happen. But it is also marked by a sense of their contingency, a point that is crucial to his last novels of historical praxis. It is this combination of certainty about their happening, and his moral confidence that they are retrograde, a kind of reluctant helplessness in face of this historical process, which makes the rejection so passionate, his sarcasm so vicious. He takes refuge, in a way, from the stage of ongoing history on which he is entirely powerless, to the stage of discourse, a world made entirely out of the referential quality of language/words, in which he can take victory away from them, a world in which he can resist history, because he is its creator. This is his world. Here he can name things into existence, and out of it. He uses this tragic sovereignty of language to avenge himself and his people on history.

In fact, two ways of this strange revenge are possible through savage satires and counterfactuals. One way is to show what has happened as a tragedy, as bad for his people; the other is to emphasize the contingency of this 'necessity', pick a suitable break-off point and create an alternative trajectory of history which gives moral comfort. This is, to use a phrase that is both so illegitimate and so apposite, Bankim's way of 'rubbing history against the grain'.[61]

Prasanna in *Kamalākānta* or her counterparts in the novels are not depicted as examples of an existing prototype of women. Women were objects of history. For all their symbolic association with the nation's power, they are also undergoing a crucial transformation. But Bankim's theory of this transition is interestingly different from the tradition of Naoroji. Naoroji represents a trend of social reflection which believes in the rationalistic transferability of social institutions and types of individual character. The criticism of un-British rule in India could begin only on the prior assumption of a possibility of British rule in his sense, which is crudely, ruling India the way Britain is ruled, establishing in reality the parallelism promised in colonial ideology. What is remarkable in Bankim is his strong disagreement with this view, with the whole apparatus of presuppositions of this sort of theory of colonial society. 'If you expect Pañcī to transform herself into an English lady (an untranslatable phrase, *bilātī mem*), you could also expect that the indigenous *śāl* tree would some day turn into an oak'.[62] It would be wrong to dismiss this as a purely culturalist argument, for the political complements of these ideas are always present, although for understandable reasons, muted, dislocated, displaced. The women, who live always near the babu but defying him, thus contain the principle of redemption for himself and his own society. And what could be more typical of the heteronomy of the Bengali intellectualist middle class (not all intellectuals; but a middle class which evidently believed and does to this day, that by simply belonging to a particular social group one becomes an intellectual and empowered to legislate solutions to the world's many evils) that it would not discover this principle of defiance, the implement to end his subjection, and to feel contempt for his only means of escape?

If we add all the different segments of Bankim's works which reflect on the question of women, they reveal a deep internal consistency. There are several places where the problem of women appears directly or under artistic or humorous disguise. Essays occasionally discuss it seriously. Indeed, the celebrated essay on equality explicitly connects the class question with the question of gender inequality, though the individual conclusions Bankim arrived at would not find favour in a more progressive age. Still, the way he generalizes the question of equality is remarkable, for it perceptively asserts that all inequalities are interconnected, and form, to use more modern language, a structure; and the inequality that human persons experience in their life world are overdetermined forms of various discrete types of inequalities.

> All human beings have the same rights—this is the essence of the doctrine of equality. I have discussed the inequality between the landlord and the tenant as an instance of the infringement of the principle of equality. Now I wish to discuss the inequality between men and women as a second example.
>
> All human beings have equal rights. Women are human beings; therefore, they must enjoy rights equal to those of men. Whatever are the activities to which men have a right, it is reasonable that women have an exactly equal right to those. Why should it be otherwise? . . .[63]

Bankim briefly discusses the conservative argument that rights are related to capabilities, and since the capabilities of men and women are naturally different, rights cannot be equal between the two sexes. This is seen as unconvincing. 'First, we reject the idea that if there are differences in nature this justifies differences in rights. This would destroy the principle of equality by its roots. For just as there are natural differences between men and women, there are exactly similar differences between Englishmen and Bengalis. The Englishman is powerful, strong, the Bengali weak, the English brave, Bengalis cowardly, the English hardy, the Bengali fragile, and so on and on. If such differences in disposition had justified differences in their rights, why do we shout so loudly at even minor differences in rights between Englishmen and Bengalis? If it is reasonable to suppose that men are masters and women are their slaves; then it is equally reasonable that the English are masters and Bengalis slaves'.[64] This is backed up by an argument really ingenious for a Hindu Bengali of the nineteenth century: 'Secondly, those areas in which there are evident differences in the rights of men and women, in these areas there are no real natural differences between them. Whatever is there, is due to the bad effects of social customs. The object of the theory of equality is to eradicate these practices'.[65]

Although all societies are male dominated, Indian society can claim a special distinction. 'The extent to which women are repressed in our society, in England and America, they do not suffer even a hundredth part of this. Our country is a land of subordination/repression. Every kind of repression, once sown in its seed, sprouts out and flourishes because of the fertility of its soil. Nowhere is the subject as subordinated to the ruler as here; nowhere is the uneducated as obedient to the literate; the extent to which Śudras and other lower castes are subordinated to the Brahmins, nowhere else are believers so subordinate to religious preachers/the clergy. Nowhere are the poor so repressed by the rich; and nowhere are women so subordinated to the will of the male'.[66] This is because inequalities do not stand separately, unaffected by the relation between different provinces of social relations. It follows therefore that 'if you are not able to provide for equality in all spheres completely, you shall not be able to provide for equality in any one field. For the principles of the doctrine of equality are closely interconnected'. For his times, this idea, that all inequalities reinforce each other, and the destruction of one requires the destruction of all—is hardly a common doctrine.

It is remarkable that when he moves to the plane of humour, the drift of Bankim's argument does not change. Often Bankim is negligently characterised as a conservative, an extraordinary judgement by any standards; for his arguments are not against reform, but against its pretentious hollowness. 'Our social reformers are somewhat more eager in establishing reputations than in analysing the processes of social development'.[67] Indeed, he is in favour of providing secular, independent arguments for the reform of women's lives instead of the derivative one that this is what has happened in the West.

> Development of men leads to the progress of society; but so does the development of women, for they constitute exactly half of this social whole. To say that the major objective of society is the improvement of one sex, and of the other only to the extent it serves the first, is an immoral idea. But the lawmakers of society, in all times, have fallen into this error . . . In fact, it is impossible to find any field of action in which license for women can be considered worse than male licentiousness. After all, both actions are equally immoral. Just as men have a natural right to a faithful wife, women have an exactly equal right to a monogamous husband. Yet, if a man transgresses this rule that is considered modernity (babuness); if a woman does so, this destroys all the joys of her world, she is regarded as the lowest of the low, and considered more untouchable than a leper. Why? Chastity of women is necessary for male happiness; but for the happiness of women sexual restraint by men is equally necessary. But of course men *are* society, women are nobody.

Despite all this discrimination, however, women in his early humour perform an aggressive and critical function. There are particularly two brief sketches at the end of LR which, though unremarkable for literary quality, illustrate this critical principle with great effect. The first, called *Appreciation of Bengali Literature* (*Bāṅglā Sāhityer Ādar*)[68] begins with an evening scene that must have been common in most Bengali households of the time: the dramatis personae include a man whose characteristics are typically indexed at some length: 'a high-priced highly educated Bengali babu'. Obviously, there is a sharply jesting use of the ambivalence of the term '*uccadarer*' which hangs between highly-priced and highly-valued, and it is a high education naturally which has given him this value/price. Typically, the interlocutor is a less-indexed person, marked only by her relation of subordination to him, 'his wife' (sardonically Sanskritized: *tasya bhāryā*).

The conversation begins at the babu's violet hour, when he returned at teatime after unspecified exertions in the office, that mysterious mechanism of law, justice and power, by which he reputedly controlled the world. He finds his wife reading a Bengali novel, which happens to be the *Viśavṛkṣa*, and asks why she persists in reading that stuff; 'why do you read that Bengali garbage (*chāi bhasma*)? It is better not to read anything', because those are 'immoral, obscene, filthy'. Since these vitriolic adjectives are in English and inaccessible to the wife, she asks for a translation. The babu takes the characteristic route of all textbook and external knowledge, he cannot break out of the circular paths of a resounding tautology: immoral is explained as that which is against morality. Since this remains equally unclear to his wife, he exclaims: 'No, no, do you know . . . now where do I get a Bengali word for it? . . . Actually, that which is not *moral* . . . there'; and further down: 'It is impossible to explain to that extent in Bengali language; but the main point is, to read Bengali books is bad'. When his wife explains that this is a book which has been translated into English, this revives the value of the work for him somewhat, but he admits that he thought it was an English work which was translated into Bengali, evidently because that is the single direction of transmission of intellectual value. It is also interesting that when his wife asks him to explain the plot of Dante to her, for she is about to read a Bengali translation, all he finds remarkable about the poet's life is that he held high appointments in Florence. He declares in the midst of the conversation: 'Bengali and such stuff are read by people of the lower class . . . We are from the polished society . . . those are written by inferior people to be read by inferior people . . . those have no value to the English [though the text uses: *sāheb loker kāche o saber dar nei*, in which *sāheb lok* can be translated equally well for Englishmen as for anglicized babus] . . . how can such things pass in polished society?'[69]

The principles of the encounter between Rāmbabu and his wife are identical.[70] Conjugal lives are constantly beset by translation problems of the Winchian sort, bringing the painful puzzlements of relativism into Bengali bedrooms. Consider the predicament of Rāmbabu, whose friend visits him on the English New Year and shakes him by the hand, and says 'how do you do?' His wife takes this, after Winch, as a mild form of physical assault, and an invitation to play a childish game of *hādudu*, inappropriate at their time of life. With great patience Rāmbabu explains to her the mysterious ways of enlightened etiquette; what his friend uttered was an advanced form of greeting, not a game; and she was misled because of the unfortunate indistinguishability of the phonetic form. But Rambabu's attempts at social explanation are not more lucid than his predecessor's. It is not *hādudu*, he says, but *hādudu*, that is how 'how do you do is pronounced.

Wife What does that mean?
Ram It means, how are you?

Wife	But how could that be? He asked you, how are you?, but you did not answer his question; you simply asked the same thing to him in return!
Ram	But that is modern civilised manners.
Wife	To say the same thing is modern manners? That means, if you ask our son, 'Why aren't you studying, you rat?', he would say to you in return, 'Why aren't you studying, you rat?' and this is regarded civilised manners?[71]

The wife implores Rāmbabu not to adopt such a civilized attitude towards her. 'You are always sick, and I have to ask you five times a day, how are you. Don't send me away with a how do you do. It does not matter if you do not adopt civilised behaviour towards me'.[72] The babu after a few more, equally unsuccessful sentences, goes in search of a lawyer to find out if Hindus are permitted by law to divorce their wives.

Clearly, the structural similarities between these two short sketches and the arguments in *Kamalākānta* are remarkable. At the most obvious level, there is the asymmetry of symbolization; the babu unsymbolically represents himself, but his interlocutor is someone who bears in herself the sign of nativism: she is a woman, who speaks the vernacular, she is the unreconstructed principle of indigenous culture. The babu knows, without reading texts, that they are bad, a miraculous power conferred by Macaulayan education. He holds anything indigenous in principle in truly Macaulayan contempt. By contrast, his wife does not take an indigenist position in culture. She is curious to understand western literature; only the babu fails to explain it to her. She does not represent an equally uncompromising rejection of western culture, and the beliefs it has got to offer. On the contrary, she represents a legitimate questioning about what could be the grounds for such a proposed paradigm shift. In these parables of incomprehension and translation-failure it is instructive to see the exact relation between the babu's powers and incapacities, because this is a sign of something much larger and more significant. Interestingly, the babu intellectual can state—but merely, *state*—rationalistic propositions. Equally significantly, he fails to give grounds for his beliefs, or indeed, to explain, i.e., to show what those beliefs consist in. Thus, these words, the magic words of rationalism, by the enunciation of which one babu recognizes another, take on something close to a ritual character.

Unfortunately, however, even women as a social group are subject to historical change. Bankim analyses the consequences of social change on the position of women with very mixed feelings. In this discussion too there is a phrase that is directly reminiscent of the celebrated temple scene in *Ānandamath*: there the discussion is about what the mother was and has become, here it is her more mundane counterparts who are the object of analysis, but the logic of the transformation is similar (*bangīya yubatīrā) ki chilen, ekṣane ki haitechen.*[73] It is true, Bankim admits, that in some ways women's taste has improved. But there are serious charges against their transformation into modernity. 'Their foremost vice is laziness. The old woman was very hardworking and was highly skilled at housework; the new woman is a great babu. Like the lotus on a sheet of water, she spends her day looking at her own beauty in the mirror in utter stillness. The whole burden of housework is left to hired maids . . . I consider it a hateful way of life if lifting expensive carpets is their only way of exerting themselves. Human beings are born in order to contribute to others' happiness. A woman who comes to the earth to loll on the bed, dress her hair in front of the mirror, roll up carpets, read 'The Exile of Sita' (Sītār vanavās), and give birth to children, who contributes to the happiness of none except her own, may be marginally superior to animals, but her womanhood is worthless. We counsel such women to rid the earth of their useless weight by applying ropes to their necks'.[174] For Bankim, clearly, the main disappointment is the conversion of women into female babus and in a society like the nineteenth-century middle-class Bengal, inevitably, when turned into babus, women's lives were bound to be more parasitic than the men's.

But characteristically, the humorous presentation of the women's question does not end here. The essay is followed by three letters allegedly written by three women. Obviously, they represent the need for lending a voice to women, because otherwise they remain ironically unrepresented in the discussions about their own condition.

Again, the point of the critique is turned instantly at the babu. The first lady essays on a symmetrical comparison between the old man and the new:

> In only one way I find the new men superior to older ones; you have learnt a bit of English. But who have you benefited by learning English? I can see that learning English has helped you learn how to be clerks, but humanity? Let me tell you what the differences between the old and new are. The old type of men looked after others' interests, you look after your own. They spoke the truth, you only what pleases. They respected their parents, you do homage only to your wife or mistress. They worshipped the gods and the brahmins, your gods are the hybrid Anglo-Indians, and your brahmins are the goldsmiths. True, they worshipped idols, you worship the bottle . . . We are idle; but you are not merely idle, you are babu. The English rulers lead you by the nose, and you follow them because you have no strength; we too lead you by the nose, and you follow because you have no intelligence.[75]

The second correspondent simply accepts the charges, but adds innocently that these are due precisely to what men have done to women and taught them: 'you are preceptors, we are disciples'. The main thrust is reserved for the last letter which suggests a role switch between men and women:

> You wish to tie us down by the bonds of religion. That does not matter. But is that extra bond necessary over and above the bond of fasts and rituals? You take the responsibility of the fasts and we would be only too willing to tie these religious bonds more tightly than ever. I wish so dearly to exchange existences with you. Before you start abusing us, take a good reckoning of the respective pleasures and pain. When we die, you would fast in our everlasting memory, you would not eat meat, you would wear rags; when you go to heaven we would savour a second conjugal life; when you are alive, you would bear children, supervise the kitchen, or ceremonies, you would draw the veil over your moustaches and with the plates over heads, do the rituals, engage in womanly humour in the *bāsar* room; your happiness will know no bounds. We would go to college with our books in our adolescence; in adult life put the turban obliquely over our coiffeurs and attend office, and make speeches in the townhall to the accompaniment of jingling ornaments . . . Stop boasting and do that. You come into the household, let us go to the office. Those who are carrying others' shoes on their heads for seven hundred years, are they men? Don't you feel ashamed to speak?[76]

By iron regularity, the babu had lost this engagement as well.

The Babu's Others: The Lower Orders

Despite the great variety of imposing women in Bankim's fictional world, *Kamalākānta*, it may appear, has a different structure. The only woman in its story is a *goālinī*, the milkwoman Prasanna, the object of Kamalākānta's verbalizing attention. Unfortunately, due to their prurient interpretation of the universe, babus read this relationship in an amorous fashion. But Prasanna is, I wish to argue, hardly negligible, and she shows features which are entirely compatible with the structure of Bankim's fictive world.

Prasanna is no ordinary woman. She condenses at least two principles of otherness to the babu at the centre of the fictive stage. She is a woman, but she also belongs, firmly, not in any oblique way, to the world of the popular. She represents the people. She represents the feminine principle I have discussed earlier. She is, like her social superiors, spontaneous, completely self-assured in her world, not assailed by any doubts about effectiveness and authenticity. Whatever she does, in her brief appearances in the story, is wonderfully and self-evidently effective. She does not, like babus, even the critical ones, have to make an effort to be what she is. Being herself comes quite naturally to her. She is wholly vernacular, and more interestingly, untouched and unimpressed by the discourse or institutional theatre of the colonial administration. Unlike Kamalākānta, when she needs to confront the judge, she does so without subterfuges, without the assistance of any subtle, hidden discourse. She is the embodiment of practical syllogism. She does, of course, reside in an unspecified village in British colonial Bengal which comes under the laws of colonial administration, but she shows a wonderful incomprehension of the laws by which she is governed and being civilized. Practically, when she loses her cow, her means of production, she unmelodramatically takes recourse to a court of law, but is generally unsentimental about the qualities of colonial justice, does not expect too much out of it, and treats it with an admirable commonsensical irreverence.

But there are ways in which she is strikingly unlike her more socially exalted counterparts. She interacts with colonial law out of an external need. Otherwise, in the general administration of her life she is presumably unaffected by colonial culture. At the cost of her illiteracy, she has defended her culture and her self-respect quite successfully. The court scene reflects this quite well. Of the entire range of characters who figure in its hilarious confusion she, and in a very different way Kamalākānta, are the only ones who do not compromise themselves, or get tied into knots by their own unmastered rationalism. The text naturally divides very unequally between the verbal Kamalākānta and the reticent and active Prasanna. She has practically no share in the verbal content of the fiction, which enhances its historical truth, for she came from a group which was similarly unrepresented in the great verbalization of nineteenth-century Bengal.

Prasanna does not have such affectations of middle-class dignity. She belongs to the folk, the popular—an entirely different world that is self-sufficient, with its rules, norms, imagery, language, even its special, freer form of abuse. This is a world that is gradually marginalized until it passes out of high literature—to the great detriment of this literature itself. Later literature would capture and reflect less and less of social space, and retreat gradually into interior of interiors. Increasingly, babus, when they condescend to look at the popular would merely see an undifferentiated blur, people designated not by proper names but by sociological functions.

Prasanna, despite her sketchiness, thus turns out to be a significant figure in the artistic and social signification in Bankim's literature. First, her relation with Kamalākānta mirrors accurately the relation between parasitic intellectuals of the upper class and the sustaining lower strata. It is not surprising that she supplies Kamalākānta with food, with nothing in return. Ordinarily babus would of course claim that what they get from the lower orders is in exchange for something. Kamalākānta is the inverted babu, endowed with a self-critical and self-revealing sight, a person who accepts and actually utters the truth of the babu world. He therefore sees the relation between Prasannamayī and himself, a parallel between the rentier orders and the lower productive strata, as one of unrequited support.

But this raises a more general question. Bankim's thought is not partial, or romantically admiring about what we now call the popular. How does the popular figure in his imagination? I shall discuss this in two sections: some of the features of his depiction of the popular in this section; and other aspects of his depiction of mass-popular actions later when we discuss his historical novels.

It appears plainly counter-intuitive to suggest in any sense a strong presence of the popular in Bankim's art. After all, Bankim's whole intellectual enterprise was the creation of a high literature in Bengali. For this literature he created a new language that did not suffer from the wooden formality of Vidyāsāgar's excessively Sanskritic prose, or from the vulgar limitedness of the babu colloquial. It was unmistakably a high literature, a literate, deliberately cultured Bengali language; and it is rather odd to suggest that there could be something strongly popular inside its deliberate cultivated accents. It is easy to ignore the existence of the popular in Bankim's fiction because these elements were not present where we usually look for them. We misjudge its *site*, its manner of existing. It existed at the level of the *forms* he used in his writing but not a form that was immediately available in terms of the deliberate strategies of narrative or literary construction. There are two ways in which this occurs in his fiction. First, the social and fictive world that is depicted has a place for the lower orders in their own terms, and they are treated as subjects, not as part of a passive social landscape or as objects of ridicule. Secondly, travesty is a form of humour that is inextricably connected to the subaltern and much of the tone of Bankim's humour has the ring of travesty, and the sound of what Bakhtin has called the laughter of a thousand years.

Later literature captures less and less of social space, and retreats gradually into the interior of interiors. Literally, the babu becomes totally narcissistic. Mental or psychological novels, and romantic poetry which a graduate male usually addresses to undergraduate females, become his paradigmatic literary norms. In these two forms of literature the literary babu could see portrayed the little tempests of his middle-class soul. A poetic heartbreak followed by an unromantic utilitarian marriage was his idea of the limits of the endurable risks of human life. Internal spaces of the novel fail to record any large or historic social conflict worth the name, occasionally offering misleading and melodramatic pictures of Bengali terrorism, expressive of a perpetual fascination of the cautious middle class for the risks of revolutionary existence.

In Bankim's novels too the babu dominates the narrative, sometimes masquerading as characters in other quite inappropriate centuries.[78] But he is always seen in relation to a whole world of people, others who are not babus: his women, his servants, his milkwoman, his vendors, the entire lower orders who constitute the market-place of his everyday. The question of what to do with these others does not figure in the theoretical consciousness of early Bengali literature, but since it is a real social problem it finds insidious ways of coming into literary representation. Calcutta is a large decorated island in the sea of illiterate, unillumined rural Bengal. Individuals from the lower orders are present inside babu households, often in large numbers as servants and other types of assistants. But their social place in this society is such that they do not deserve literary speech or subjectness. Bankim shrewdly observes the connection between cultural patterns and work relationships. Productive organization within the middle-class household changes. Women become educated; unfortunately, in his view, they become babus as they too withdraw from housework; and the households of urban upper classes are invaded by an army of silent menials. Gradually the work-based distinction between the babu and his wife changes too, to her detriment. Formerly, the babu symbolized parasitism; by contrast his wife was engaged in useful work at home, which was, apart from other partial explanations, the reason for her invincibility on occasions of domestic skirmish. With changing work regimes within the Bengali home, the wife comes to acquire a new kind of workless leisure, and she turns into a greater parasite than her husband, for internal productive work is given over to a retinue of servants, and external 'productive' work—the exertions involved in the earning of a primarily rentier income—was for social reasons monopolized by the husband. Upper-class women fall quickly into a leisured heteronomous existence which was historically quite novel, probably not enjoyed (or suffered) by anyone except royal concubines. Until the Bengali woman breaks out of the confines of the home to the great liberation of the office job, she is condemned to a life of decorative superfluity.

In this new urban world, the lower orders essentially mean servants and other dependents, held in various forms of the relation of subalternity. Interestingly, by contrast, they play an important role in Bankim's fiction. There are some complex figures who are not easy to describe in these terms, like Bimalā[79] or Indirā in her fallen state in the Calcutta household. Perhaps this is a technique of double significance which as much debases the particular individual as raises the state of servanthood into a kind of salutary ambiguity. More authentic subalterns (servants)—Digvijaya, Girijāyā,[80] Rangalāl,[81] Māniklāl,[82] Rāmcaran[83]—constitute integral parts of the social world. Bankim's stories show, often, the dependence of the protagonists on their subalterns, and their parasitism. In subsequent literature, by and large, if servants do enter, they are hardly more than shadows. They never ever speak, are hardly ever part of the proceedings, except as instrumentalities. Speech here is a major sign, a mark of the autonomy, of the existence of a subject. Denial of speech consequently debases them into a mute objectivity in a world in which the babu—garrulous and fractious—alone has the right to subjectivity. Even the *bāuls* and *bairāgīs* in Tagore are impeccably babu in their language and cultural taste.

It is of course possible, and I think correct, to suggest that this place of the lower orders as speaking subjects in Bankim could be a dual feature. Part of its inspiration could in fact be drawn directly from European literature. In the babus' contact with the West, literature played a major part. This was not only due to the natural curiosity of the new middle-class intelligentsia about western literature as a mirror of western society. Literature portrayed two features of western culture which they considered significant. First, novels, because they were mainly realistic, provided a reliable picture of the mechanisms of social relations. Secondly, the novel was also an expression of a rationalistic literary form; it was part of rationalism that the representation of society in art was made in the analytical form of the novel. The new Bengali intelligentsia therefore went into western literature with great enthusiasm, often to a complete neglect of their own literary heritage which came to be regarded, superstitiously, as unreadably vulgar.

Exposure to English literature involved acceptance of the literary norms of criticism and construction. Contemporary literary forms could be made sense of only through their history, and Bengali intellectuals' reading ranged equally over the classical and the modern literary forms. Renaissance literature therefore constituted for the Bengali babu the highest classical canons of literary excellence. I am not concerned here with its general influence, but only with the question of how it affected the depiction of the lower orders. Evidently, the internal social composition in the high literature of Europe had a dual structure. In Shakespeare, as much as in Molière, the gentleman is nearly always accompanied by a man-in-waiting, a servant who is also a friend, a Sganarelle to a Don Juan. Obviously, this is a double technique. Only partly is this a mere depiction of the social rituals of noble society; it is narratively far more significant as the use of a different eye, and a different voice, which plays against the eye and the voice of the master. By the nineteenth century in European literature, such closeness and dependence between the gentleman master and his servant is already rare, and replaced by a formalized bourgeois social distance which does not allow the lower order persons to play a speaking role in the story. They go away to live primarily in the murky margins of polite society as servants or criminals. Occasionally, they come to have a literature of their own, as in Zola.

This could certainly be one possible literary source of the lower order figures in Bankim's novels. But it also has another unmistakable function within these narratives. They contain a reflection of the social and symbolic order at that historical point in Bengali society. He lived at a time when modern class distances, though already instituted and beginning to be codified, had not crystallized as they would in immediately following decades. However, I think it would be an error to put their ability to speak down only to the survival of European canonical forms, or the state of affairs in the Bengali family circle. It is because Bankim has a dual, contradictory relation to the culture that the babu world was creating for his country, which makes him turn repeatedly to a lower order (or perhaps better, a non-babu) voice. It is not its frequency that matters, but its symbolic significance; not how many times this happens, but when. It is in people of these lower orders that Bankim finds a directness, authenticity and unassimilated quality that can act as a foil to the inauthenticity of the colonial middle class.

What I wish to suggest is that in Bankim's art the social world seen in terms of people who are considered admissible within the boundaries of this respected, enchanted, sacred space, is wider and consequently more realistic. Of course the babu dominates this world, both inside fiction as well as outside it, but, because his world has not been so irreversibly formed, so strongly and irreversibly structured with the stamp of his dominance, he still lives in close neighbourliness with the lower orders of people. Servants are almost always on attendance on their babu principals, not silently in demeaning postures of servility, but with the ability to speak. And this ability immediately confers on them a different identity and vision, a distinct social sight. This sight is not different simply in terms of social perspectivality, but also a different kind of subtlety and perceptiveness. Not only does the arrangement of the social world appear differently to them; because they are free of typical higher order inhibitions they are able to see a whole number of things which their superiors are either not able to see, or admit they have seen. In European literature classical drama is the great illustration of this technique, particularly Shakespeare and Molière, in whose work the underside of things figures constantly and comes up for explicit discussion. Occasionally, lower characters are used to observe reality in a related but different register. While main protagonists would often register a complex and vague impression like beauty, they might often redescribe the same person in terms of detailed anatomical observation. The bodily attractiveness of people, the materiality of the universe, and the slovenliness of social relations figure largely in their observation.

Their presence in the narrative is not just a literary excuse, not just a matter of dramatic empiricism—a technique of physical portrayal to make the setting credible. They are not just milieux; they are vital to both the structure of narrative and speech internal to the drama or the novel. For the point of the drama and the novel is to reveal the truth of the world. They bring a counterpointing commentary, a counterpointing experience of relations and objects, a greater earthiness of speech and observation to balance, to complement, to fill out the absence and the inabilities of their superiors, who are encumbered and deceived by their own structures of censorship in preception of the world. Servants' voices, or of lower class figures represent voices and visions from outside the region of the repressed. Surely, such literature does not amount to a conscious celebration of the culture of the people—of the folk, the grotesque, the bodily, the licentious as does the Rabelaisian moment in European literature.[84] Still, it shows an interesting equipoise, at least a critical sense of this advancing repression, the closure and disqualification of whole branches of existence, whole terrains, on the ground that they are unfit for aesthetic, literary representation. The classical vision shows a strong sense that this excluded world is part of the world, and that without some representation of this, the picture of the world would remain untrue. Between the idealism of the master and the materialism of his servants, the literary world achieves its fictional truth.

In the play of formal elements in Bankim's humour there is a strange and often unremarked mixture. By form I refer to two aspects which are analytically or theoretically distinct. One aspect of formal artifice consists in the use of language to create literary tropes. In terms of tropes, Bankim's favourite is the ironical praise (or praise abuse), a piece of *alaṃkāra* taken straight out of classical Sanskrit rhetoric. *Alaṃkāric* play of this type can obviously be appreciated only by the highly literate, who can understand the rich vocabulary of classical Sanskrit, and the still greater resources it gives itself by rules of grammatical manipulation and *samāsa* and *sandhi*. Sometimes, as in case of the babu *stotra*, it creates its humour by use of a high language for a low occasion, giving the babu's less exalted pursuits the ceremonial description of high ritual. None of these features can be called popular. But form does not merely refer to stylistic, but also to the philosophical tone, or way of relating to the world. In writing, this is the equivalent of what tone is in spoken utterance. And Bankim's manner of relating to the world all through is *travestic*, and travesty arises out of an urge to commit social sacrilege. It emerges through a popular urge to desecrate the high rituals of a constituted society which the critics are powerless to challenge or transform. Those who lack the means of power to recreate the world according to their preferences, or answer everyday insults, take their revenge through travesty, the typical weapon of the powerless. Bankim's art, if we read it in this way, achieves a curious mixture of the use of humour and satire by the traditions of the élite and the popular, while the language in which it speaks is élite, the way the world is brought into representation inside it is travestic and popular.

This is one of several ways in which Bankim's art stands still fairly close to earlier traditional forms of literary entertainment. Literature is still pre-eminently a form of play—its play with language, with its audible pleasures and idiosyncracies, the enjoyment of the whimsical play of sounds. Subsequent Bengali literature would look upon this with slight contempt, as unworthy of serious and subtle writers, and the enjoyment of language would turn more on a play of meanings than of sounds. This also reveals itself in the quality of its humour—the rawness, directness and non-stylized character. Bankim's novels, or even his essays, if not like a market-place in its noise, laughter, lack of formal structure and severity, its unstructured enjoyment of life like Bakhtin's *Rabelais*, is at least a much wider, open world. By the time we come to Tagore, this laughter, this openness, this vitality has gone out of the novel form. The history of Bengali literature afterwards is a history of a great narrowing of the social world.

The Narrowing of the World

The world becomes narrower because of, and in exact measure with, the babu's conquest of it. Literature, or the world to be presented in it, runs increasingly through a double filter: a filter that is aesthetic and at the same time surreptitiously social. Nowhere is this reflected more than in the internal complexities of the language. Certainly, in a sense, the language becomes more refined, capable of handling greater shades of situational and emotional tones, more developed as a literary vehicle. At the same time it becomes more one-dimensional. It becomes exclusively the language of the babu, the language he would have considered refined, proper, fit for enjoyment. He imposes in other words the history of his class as a restrictive format on the language of his culture. Other languages are marked by a shamefacedness, and gradually disappear. Other languages, still present in Bankim, are most severely filtered and censored out of it. Consider in contrast to this, Bankim's manifesto: 'You must express clearly what you wish to say: you must say all that you have to say; use whatever language is required for that purpose—English, Persian, Arabic, Sanskrit, rustic, savage, whatever language, take recourse to anything except the obscene'.[85] This linguistic programme would become entirely unacceptable within two decades.

To relate the actual use of language and choice of words of a writer to the social world is a risky, if not misleading activity. Still, it is an intuitive truth that there are real and immediately perceived frontiers across the spoken and used language.[86] In the language that a writer uses, there is sometimes a deliberate strategy at work. Indeed, since the writer's production is related to words, no writer can have a purely unthought-out relation to the language in which he writes and the language which gets represented in his writing. Even though he does not have a wholly deliberate strategy, either through design or half-indifference, a writer lets some parts of language and some speakers and some modes of speaking to figure in his work. This could be used to discern his relation to the social world.

Bankim's language is not only wonderfully innovative, but also strikingly open. The idea of openness means simply that it lets the Bengali world, before it becomes ordered and arranged according to a predominantly babu taste, to be represented inside literature. It is a linguistic world full of tensions, of alternative possibilities. One cannot say that Bankim uses the language of the market-place. But one could use the market-place as a metaphor. If one wished to see the whole social world, to listen to it, the market was the proper place to do so. A market is the great social crossroad, where different classes come and interact and can be themselves. It is in the market that one can hope to hear all the accents, all the social dictions of that society's language. Markets are therefore microcosms of the social universe as nothing else can be. Occasionally, one hears something like a market-place effect in Bankim's writing. Of course these languages carry their dense social markers. Bhadralok speak Bengali differently from the lower classes. Bhadralok speak differently when they speak to other bhadralok and when they are speaking to their inferiors. Within the bhadralok household the women represent a linguistic opposition. Later, this is replaced by a dull uniformity of bhadralok language. There could be great divergence of political opinion in a drawing-room conversation, but these would all be expressed in the same drawing-room dialect. Depending, therefore, on what you hear or set out to hear, the conversation could be varied or uniform. To someone like Shaw's Higgins,[87] it would be unbearably flat. He would say we are doing the wrong kind of listening.

To turn this around, a piece of fiction that is moderate in terms of the opinions explicitly expressed could be astonishingly varied and popular in terms of the sounds it allows entry within its fictive circle. Modern literary listening in Bengali, in this sense, is impoverished and one-dimensional. We have been trained now to listen to meanings, and somehow do not hear the sounds, reducing language to only its semantic register and entirely obliterating the phonetic one.

Language for Bankim in any case is a theatre of war.[88] English culture, in its imitated babu form, is always the addressee and the adversary. And in this, Bankim seems to offer a solution to the problem of the linguistic crisis of Bengali which went right against the predominant thinking offered by the rest of Bengali culture of this time. Comparison with the English language threw the Bengali language into a crisis; for it was clear that it must, in order to compel attention, offer resources of expression which are comparable and equally versatile. But the standard solution, which was historically accepted, was to integrate uses of English syntax and vocabulary, and to affect an increasingly refined, intellectualized style. Elements of the popular were generally excised from its ideals. Bankim's solution is to take the entire world of spoken and used language and use it against the invading English syntax. Because of that theory later literary language becomes flatter in terms of social tonality, more impersonal, more formally equal.

Bankim's artistic practice therefore shows a certain adequacy to his aesthetic theory. *Kamalākānta* shows this aesthetic quite clearly even through its tone of half-seriousness, an aesthetic ecumenism which uses the historical availability of several traditions to fashion a creative synthesis, rather than an impoverished westernized solution. Unlike later authors, he allows these different traditions to speak in their different voices, and he has the amazing, practically unique, ability to select and mix their strengths.[89] Later these diverse tendencies are excluded, censored, debarred. In the culture of Tagore, for instance, the high classical tone derived from Indian influences mix easily with high western aesthetic in a self-conscious and deliberate theoretical construct. But the element of the popular, the carnivalesque, the market-like is missing. Of course in a certain way, the later culture is more refined and cultivated, but it contains an erasure which is worth analysing.

The End of Irony

Kamalākānta ends characteristically. I deliberately say characteristically, because it displays the almost casual nature of the ending common to much of Bankim's non-fictional writing. It is an end that hangs loose in terms of form, an end that lacks the finality of a denouement, an ending after which the narrative can never go on. Fiction and non-fictional pieces have very different conclusions in Bankim's work. Endings in his humorous writings show a less deliberate quality. Surely, both the humorous essays and serious fiction are commentaries upon the world, but they accomplish this comment in very different ways. Novels, through their deliberate form, make philosophic statements about the nature of the world and the destiny of men. Humour comments on the irrationalities of the everyday, in a manner which also has a bit of the everyday in it; it has the everydayness of form in its relative lack of boundaries. It can be dropped for the time, and can be resumed in a manner that is less theatrical. Yet I feel this formlessness carries a symbolic effect that is hardly less powerful than the Gothic endings of his better novels. Probably satire is left like that because they provide a commentary on his life, his society, the entire ongoing business of his world; and perhaps there is an untheorized adequacy between the unconcluded nature of this humorous comment and the unconcluded nature of the world. The colonial everyday experience, the sound and fury of the Calcutta babu's intellectual life, the inconsequentiality of his existence was a comedy that was still going on, and therefore it would perhaps be inappropriate if its humorous accompaniment gave a final conclusion to itself. The text of *Kamalākānta* thus shows a peculiar recursiveness; it ends after the *daptar*, then Bankim adds on the letters, then again the deposition, and finally, the appendix. It ends several times, without actually ending. Bankim always finds some excuse, some peculiar continuatory form, to justify a re-opening of his hostility with this world.

Despite its provisionality, *Kamalākānta*, alas, has an end. In this last scene, his interlocutor is an imaginary speaking cockatoo. It says things that are rather similar to the ideology constructed and repeated by the colonial administrator and his ally, the rationalist intellectual; but this is offset by the bird's pleasingly disconcerting habit of declaring *platitude*, as a sort of authoritative commentary on his own discourse. There is no way of knowing whether this is said about what others say, or what he says himself, or whether this enunciates a deeper and more fundamental theory of discourse under colonialism; i.e., it is only platitudes which have a title to be stated ceremonially.

At the end of the piece, Kamalākānta notices a crowd of small creatures who in his blurred vision resemble ants. But the bird explains: 'True, these creatures are very small, like ants; they look like ants too, but they are not ants really. They are called Bengalis. Look, a small drop of milk just trickled down from my perch; and the Bengalis fell over it and started fighting with each other for a share.[90] They do indeed live on stray drops of milk dripping from my perch occasionally; and still you have the nerve to say that I am not a benefactor to them?'[91] When Kamalākānta undergoes an awakening from his opium-induced stupor he finds himself face to face with Prasanna, his foil, witness to his secret helplessness. Kamalākānta now finds a heap of ants on the ground. His delusion, the overlap between literal and metaphorical reality, between Bengalis and ants, his second sight which enables him to see as no one else can, has not quite left him. He still prophetically confuses between the ants and the Bengalis and asks Prasanna to take a broom and 'sweep these Bengalis away' (*oi jhāntā-gāchtā diā bangajagulāke jhāntāiyā pheliā dāo to*).[92] And the woman does exactly that (*goālinī māgī tāhāi karila*). This is the last symbolic gesture on which Kamalākānta's dream ends, or you may say, begins. Prasanna, the repository of the qualities of the people, untutored, unspoilt, undegraded, like his other women, becomes the figure of history. She does in the final line of the text, what the author and the dreamer had been trying to do all through by his humour; this great act of cancellation, denying the reality of the present life of the Bengalis. She simply negates, cancels, refutes them. She sweeps them away. It is only on this condition, within this negated, cleaned, new space that a new country, a new people—his native land—can begin.

The self-ironical discourse in *Kamalākānta* had reached a kind of internal limit. After reaching that limit, to move forward, Bankim has to invent a different people by inventing a different history for them. This account must be different from what actually happened, a terrain of *imaginary* history. Kamalākānta thus represents a break in the evolution of Bankim's thought. After this, the discourse of the ironizing self would not be enough. It would be necessary to invent another self which would be able to break out of the prison of history—not in humour, but in truth.

3

The Myth of Praxis

Construction of the Figure of Kṛṣṇa in *Kṛṣṇacaritra*[1]

My relation to religion is no merely negative relation. It is a critical relation. I only separate the true from the false, though it cannot be denied that the truth thus separated from falsehood is always a new truth essentially different from the old.

Feuerbach[2]

Classicism is not a matter of antiquity, but of form. Ideas become classical not through their continuity in real time, but in an imagined time, substantially and interestingly independent of real history. A figure is turned classic not by a period of long uninterrupted existence, which allows a constant process of refinement that pares away everything except the most fundamental attributes, the most essential outlines, and achieves what appears to be an unsurpassable economy and ideality of form.[3] A classic figure, curiously, need not be constructed that way at all. It can be constructed quite suddenly, by the requirement of historical circumstances in which an intelligentsia might invent a classic tradition for itself in a matter of a few years. What matters is not its existence and continuity in real history but its appearance of doing so. Sometimes, in admittedly the rarest of cases, it could be constructed by a single text. A construction of this kind is what Bankimchandra's *Kṛṣṇacaritra* (*KC*) seeks to do. It is a commonplace in the history of ideas that the past, at least the past which lives in popular imagination and acts upon history through popular behaviour, is a construct of the present, the past that the present would have liked to have had, which it considers appropriate for itself, its self-created lineage. It is the past out of a range of possible ones which would give to the present and its self-image the greatest degree of coherence, dignity and justifiability. In Bankim's *KC* we find this imaginary past in a rare condition of plasticity, literally in the process of being formed.

Bankim's *Kṛṣṇa* is the Almighty's latest hermeneutic form, quite different from this god's commonly known figure in Vidyāpati, Caṇḍidās, Caitanya, Jayadeva or the *Purānas*.[4] But there is an even greater, deeper difference. This is the first *Kṛṣṇa* sought to be created by a genuine historical argument, that is not poetic, which self-consciously follows a method quite unlike the unrigorous procedure of narration of the *Purānas*. It does not claim self-evident correctness for its own interpretation of *Kṛṣṇa's* life-text, does not pretend that the 'text' of this life, narrated by a series of diverse narrative texts, is whole or internally consistent.[5] One could argue that in Bankim's book, precisely these two ideas of textuality (or ways of being text) meet in a subtle and interesting conflict. Bankim brings in and acts on a western, or at least West-derived theory of textuality, but the object of his analysis and attention is a text of traditional Indian religion; and this tradition implicitly contains a very different theory of what a text is, what its criterial properties are, how it is supposed to exist, what it is meant to do. Unlike texts in the Hindu tradition, Bankim's construction does not alter the earlier line of texts subtly, surreptitiously, by altering the totality of the image, gradually shifting its centre of narrative gravity.[6] Traditionally this was accomplished by adding new episodes to an old story, but in such a way that after a point the additions altered its truth. Bankim's work takes its historicity very seriously, differentiates sharply between itself and other texts, and brings into play the idea that one of these many images of Kṛṣṇa must be true and the others false. Earlier texts like the *Purānas* work on the theory that one text does not oppose or contradict another, but worked, to invoke the metaphor Bankim uses so angrily, as 'deposits of interpretation', like lamination on a vase, darkening or lightening, altering its colour. On that theory, one reading, though different, is not in competition with others; all constructions somehow contribute to an idea of what god is.

On Textuality

The Indian and western intellectual traditions seem to be quite different not only in the content, but in what could be called, *faute de mieux*, the *form* of their texts, i.e., not only in what the texts say, but in what they are. The main point is not that Indian tradition is preliterate, that it does not know writing. Obviously, it does know writing, but its attitude towards writing, I shall call this its theory, is different. Much of the internal cultural transmission of significant ideas in the Hindu tradition is oral; but this is obviously different from the orality of a culture which does not know the apparatus of writing. Its peculiar attitude is perhaps better expressed as the orality of a culture which knows writing but mistrusts it. It follows that its attitude towards the text is different: texts are a weave of words into something aspiring to a peculiar objectivity.

Between the text and the authorial signature (or authorial individuality) the Indian tradition sets up a distinctive relation. In the western tradition, in comparably classical antiquity, and increasingly in the period after the Renaissance, a text bears an inextricable relation with the author's person. Although there are similarities in form between hymns and stotras, medieval Christian texts too are marked by their individual authors.[7] A text is the author's creation, something for which the author is an efficient and irreplaceable cause. A text is unthinkable without its author, for he supplies it not merely with accidental attributes of style, form, internal structure of argument, but its entire way of looking at and making sense of the world. Texts are therefore traces or monuments of individual ways of seeing and constructing the world. So, although the author admittedly creates this world within his text through public or social resources, nonetheless, it was the individual's way of looking and making sense which made it valuable, irreplaceable, which gave it an appointed and ineradicable place in its historical universe of discourse.

The function of the text in the Indian tradition appears to be significantly different. Of course authors who somehow survive the erasure of time, like Manu or Pāṇini, wrote what were never intended to be misclassified as a celebration of the individual's way of seeing, whose value lay in the peculiarity and the indelible mark of his personal view. Rather the talent of the author consisted in a kind of erasure of the individual, not in taking an interesting individual line which contested, competed with others or replaced them by virtue of its greater truth value, but precisely in his ability to make himself transparent and glassy, and his text point-of-view-less as it were.[8] Texts that were created by this kind of peculiar authorship acquired an objectivity, but in a sense different from the usual meaning of the term. There is, it seems, a deeper theory working behind this culture which claims a special objectivity for the mark, the trace, the inscription. It is a general quality of inscriptions to objectify, to render an idea timeless, above subjectivity; and writing had this quality irrespective of whether it was an inscription on a brittle piece of paper, on stone, or the very paradoxical medium of collective memory. It was not the locus, the medium, that was important but the mode—inscription itself. Only things which rose above the contingency of the subjective were entitled to this significance. Consequently, Asoka's inscriptions on stone and Pāṇini's on less durable material, shared a common aspiration towards a moderate eternity. This is why texts are so few, and so extremely important.

In the Indian tradition, the text not only hides the individual (which it does more or less in every civilization: we know very little of Plato except his books), it erases him. Perhaps it could be put still more strongly—it was only someone who could erase himself who qualified to be a writer of such frightfully objective of texts. The pretence of a collective mind in the act of inscribing is exceptionally strong. The author is not the text's creator, but its instrument. He is almost an amanuensis of a structure of ideas enriched, perfected, economized and sharpened by an impersonal tradition. This is what leads to the dual and interconnected forms of the *sūtra* and *bhāṣya*, this combination of extreme compression and extreme elaboration, of economy and prolixity, carrying to their logical limits both functions of words.

Textual hermeneutics stretches therefore from the level of overarching structures of philosophical meaning, to the sense, to the reference, to the literality of what is said—a hermeneutic detail necessary to bring under control the discursive and mnemonic tradition of the *sūtra*.[9] Textuality, fixing some idea into the form of a text, is therefore always two-valued and ambiguous. They objectified and fixed utterances, meanings, gestures or memorable acts. This fixity however was nothing more than an assurance of constant remembrance—that the object would recur endlessly in relevant discourse, but it could not ensure that it would mean the same thing or be remembered in exactly the same way. Giving fixity to something also simultaneously invited constant interpretative change.

Religious texts which celebrated and narrated the life story of *avatāras* were texts about texts in a manner of speaking. These rendered into the objectivity of written words the message of the great figure, or inscribed on collective memory the other, more fundamental sacred text of the saint's or hero's life. The *Gītā* is thus an inscription within an inscription. Acts of Kṛṣṇa inscribed the meaning of goodness in a life-text; but the *Gītā* is also a spoken text inside this life-text which explicates unmistakably what Kṛṣṇa's life means. It is a commentary on his life which is part of his life. Rāma's life by contrast is relatively non-discursive.

The way such lives are lived or executed, is similar to the process by which art is created; in its deliberateness, its choice of means, its search for a place of inscription which is indelible and changeful, as the word or the collective memory. Some of the major features of the Indian tradition are related to this problem: they arise not out of what is written, but what is written on, the strange tablet of memory of collective imagination which effaces and retains, which cannot retain an image without changing it constantly. The life of a great figure is a text of pure interconnected acts. It is unlike the weak texts made out of words. These lives convinced men not by the abstractness of moral concepts, compounded by the symbolic abstractness of words, but by the resplendent directness, the incontrovertibility of the act itself, by dispensing in a sense with secondary interpretations. Such non-discursive texts are not a novelty of the Indian tradition alone. Practically all religions, or movements which had to convey practical ideas to large uneducated masses, produced such figures. The life of Jesus is a text as much as the life of Caitanya was. The life of Kṛṣṇa is also a text of this sort.

Exegetic Principles

Hindu theology has a considerable literature on the intricacies of incarnation; there is therefore a good deal of discussion which analyses in detail the exact nature of an avatāra, i.e., the exact sense in which he represents God. All avatāras are of course representations of the absolute, but they are so to different degrees, and in different ways; avatāras often represent some aspects or some proportion of Him, 'part' of a god, or his *ansa.*[10] The general argument about avatāras is fairly straightforward and well known. If God is absolute and infinite, and if at the same time He is the most preferable object of knowledge, a problem arises. Human minds are after all finite, and this leads to the familiar problem in all philosophic religions—how can finite minds know an infinite object?[11] The Hindu tradition chooses a simple answer (though some would perhaps point out, plausibly, that it is not an answer to exactly the same question). It says that the Absolute wishes to be known by others, and for the *vaisnava* tradition, by Himself, and He simply (and economically) arranges for this by making available a representation of His self in a finite form, an occasional avatāra. Apparently, this is the Absolute's answer to the famous puzzlements of indeterminacy of interpretation. In His fulness, His absoluteness, He is unknowable. It would be a pity of cosmic proportions if this infinitely joyous presence in reality is not perceived. That is why God seeks adequate representations, metonymic reminders of His powers, or of what He actually is.

Among the avatāras themselves, Kṛṣṇa is in a class by himself. For he is a metaphor of himself, if this idea makes any sense. His incarnation is quite often put on a different footing from all other avatāras of Viṣṇu. Do the ten avatāras represent a relativistic scheme, that is, is there an internal principle of adequation of the avatāra to the nature of the calamity he helps the world overcome? Who else but a gigantic fish could carry out the tasks he had during the *pralaya*? Who else, given Brahmā's thoughtless gift of near immortality to the evil Hiraṇyakaśipu, could have destroyed him except the logically uninvalidated form of the *nṛsimha*?[12] Later, such peculiarities of the crises and their specific deliverers and the evident animism implicit in the early avatāra figures give way to more anthropomorphic forms.[13] Surely the adequacy of Rāma to his situation is not of the same physical kind? Finally, there is an argument in the vaiṣṇava tradition that Kṛṣṇa has a different status from all other avatāras: while they were partial representations or segments of the Absolute, Kṛṣṇa is a representation in full (which again raises the question if he could be called a representation at all). While others were avatāras of Viṣṇu, Kṛṣṇa is He himself. Is He an avatāra on the same level as, say, Kalki or Rāma? Or is He his unmetaphorical self? Jayadeva, perhaps on such philosophical grounds, composed an intriguing *daṣāvatāra* invocation in which the usual avatāra Kṛṣṇa is quietly taken away from the series, and replaced, oddly, uncharacteristically, and I think beautifully, by the figure of the Buddha:

> nindasi yagnavidherahaha śrutijātaṃ
> sadayahṛdaya darśita paśughātaṃ
> keśavadhṛta buddhaśarīra
> jaya jagadīśa hare.[14]

Bankim does not get involved in such conventional debates of vaiṣṇava theological doctrine, understandably I suppose, because his point was not theological but political. He states, blandly, that he believes that Kṛṣṇa is God himself (*Kṛṣṇastu bhagavān swayam*), without showing the grounds or details of his belief. There was another problem—partly theological, partly philosophical, which he had to address at great length.

Although vaiṣṇavas could agree with Bankim about the direct divinity of Kṛṣṇa, this did not settle all disputes. For the main purpose of God's coming to earth is to emphasize the possible universality of good life among men; and there are at least two theories in the Hindu tradition about this universality. God and men could be linked universally through two different types of argument. Two attributes of man are considered universal, and therefore particularly suited to be metaphors of the divine, although their theological implications were significantly different. These two were the universalities of love, or more narrowly of sexuality, and of reason. The Kṛṣṇa cult that was available to Bankim through the powerful vaiṣṇava tradition of Bengal, was predominantly a cult of love or sexuality.[15] In both theories it was maintained that the touch of God or the way to Him must be universal in principle i.e., if it was not already present in all men, it should be attainable by all. To identify God with reason (*gnāna*) or to say that reason is the path of attaining Him, is to immediately set up a hierarchy; for although men share rationality as a common attribute, as European rationalists never tired of repeating, they do not have it to the same degree. The trouble is that reason or its accomplishments are comparable. By contrast, to identify the universality of *bhakti* with the universal availability of sexual pleasure was to make the doctrine of the availability of God more open. It was the most universally available and acclaimed of pleasures (*ānanda*) and introduced an intriguing element of incommensurability into the matter. Eventually, Bankim's

Kṛṣṇa would be very different from the Kṛṣṇa figure available through the vaiṣṇava cults of Bengal in which this principle of sexuality predominates—a Kṛṣṇa too familiar, too folkish for his liking, too obviously invested, though lovingly, with the frailties of a mundane eroticism, a tradition in which eroticism had lost its original metaphorical quality.[16] Eroticism turned gradually literal, and poets especially revelled in their descriptions of its suspiciously material sounding raptures.[17] This was a process begun, in Bankim's view, by the sensual materiality of Jayadeva's great epic, and carried on by lesser poets of Bengal.[18] Curiously, Bankim does not observe that Jayadeva's poem also has a classicizing quality about it, although it was a classicization of eroticism. Its form shows this, for it depicts essentially a cycle of timelessness.[19] The *Gītagovinda* is not a story in the usual sense. Ordinary stories hold our interest because of their contingency, the accidental quality of the narrative they portray: we wish to know what happened next. In a 'story' of the classical form however, what 'happens' does not allow for such idle curiosity; for we already know what is going to happen in a quotidian sense. Separation is followed by reunion. An initial infatuation can be riven by philanderings and crossed by other infatuations. To be deep and to attain its truth, love must be crossed and enriched by suffering.[20] In a strict sense the poem does not consummate the curiosity of the prurient. The constant and endless recursiveness of descriptions shows a typical feature of the classic style; to attain the unattainably perfect description—the ultimate triumph of language over the elusiveness, indescribability of reality.

The classicism of Kṛṣṇa that Bankim would try to construct is very different from it all. At least on two points the difference is sharply evident. First, he mistrusts sensuousness, even as allegory. The Kṛṣṇa figure cannot be represented sensuously because of the thinness of the division between classicism and sensualism, the presence of the insidiously obscene behind the evidently classical. Bankim mistrusts still more the tension and instability in the figure of Rādhā whose metaphorical quality can dissolve suddenly into an unmetaphorical and unmediated riot of sensuality. Secondly, as a result, this new classical figure has to be rational and intellectual. He is seeking a Kṛṣṇa who is not an object of a supplicatory form of bhakti,[21] but a symbol of praxis, and before we can go into praxis we must cognitively master the world. The Kṛṣṇa of his making cannot be the Kṛṣṇa of the touch or of the affective, it has to be a figure of pre-eminent intellect.[22] We can see now the full complexity and contradiction within Bankim's intellectual project. At bottom, there is the text of a divine life; for the life of Kṛṣṇa is lived precisely to be a text.[23] Kṛṣṇa's life is meant to fix the pattern of ideal human behaviour, of human personality developed in its totality, to prevent its principles from becoming vague and sullied through doctrinal contention. That text of his life, however, to be saved from fading out of a failure of mnemonic devices or through the insidious decay of interpretation of words—because of the indeterminacy that comes in when the directness and concreteness of acts are given over to the abstractness of discourses—has to be fixed in a medium of written texts. These texts are the epic and the *Purānas*, but the *Purānas* in trying to fulfil this purpose actually undermine it. Bankim's own text is about these intervening texts and mainly about what to do with their cheerful inconsistency. It tries to dig through the 'distortions' of Kṛṣṇa's life in these texts to the life 'itself'. To do that it has to displace and reject the texts lying in the middle. Paradoxically, and faultily, he wishes to recover the life story of Kṛṣṇa not as a mythical or fictive object, but literally as an object of history.

If Kṛṣṇa is God himself what could be more appropriate or uplifting than a constant incantation of his name? If the everyday life of Bengal is already inscribed by His name (though perhaps absent-mindedly) should this not be sufficient for the origins of a regenerative process? To Bankim that seems unlikely, for the name of Kṛṣṇa is remembered falsely, and his life is perversely interpreted.[24] Consider the events of his life: if the folk cult is to be believed, Kṛṣṇa was a thief in his childhood, an adulterer in adolescence, an imposter in maturity. Are these acts consistent with the nature of the divine? A possible argument, with which Bankim would have agreed, could be that people constructed the character of their gods according to their own; and the popular figure of Kṛṣṇa is more an approximation of the general character of Bengalis than of the nature of God. In a conventional way, one could ascribe these acts to His *līlā*; but this would still seem perversely inclined towards frivolity if not actual baseness. Would the Divine Being, when He adopts human form to instruct humanity in ideal behaviour, engage in such paradoxical consistency of petty to medium-sized vice (*pāpācaran*)?[25] His western education, Bankim claims, has reinforced his belief in the direct divinity of Kṛṣṇa; but obviously this requires detailed disputation and judgement, for it is only a reconstructed Kṛṣṇa to whom such rational respect could properly be rendered. 'At the present time of intense religious debates, it is necessary to analyse the figure of Kṛṣṇa in detail. If we have to retain what is old, we must find out if it has something worth preserving; on the other hand, even if we have to abolish the old, we must analyse Kṛṣṇa because you cannot abolish the old without abolishing him'.[26] Religious debates are thus the front for something else, something deeper, more disturbing and undeniable and Bankim reads it quite correctly—debates about religion, about knowledge, about aesthetics, are all inserted into an unpronounced debate about colonialism.

Bankim thus regards Kṛṣṇa's life as a 'master narrative' in the purest sense,[27] a narrative construction that, precisely through the exaggeration provided by fictionality, reveals and exemplifies the central organizing principles of a culture. Narratives of this kind are significant because they are not the creation of a single author; they contain the process of a slow mosaic of a story through long-term refinement and presentation of a culture ideal of humanity.[28] It alienates into an object, to use Feuerbach's phrase, and thus objectifies what it considers to be the essence of human conduct ordinarily unattainable within the ambit of ordinary human lives.[29] Had this story been a single individual's creation, it would have been idiosyncratic and fallible. But the figure of Kṛṣṇa came to be constructed through something like a massive collective sculpture, through a long series of texts and textual practices like retelling. Its outlines, its emphases and inflections were thought, corrected, rethought through a collective historical process that is not entirely unmystical. This is what lends this peculiar and intense seriousness to a judgement on Kṛṣṇa's character. For what is being judged is not a poet, not a sect, not a religion, but the self-presentation of Hindu civilization, what Hinduism has thought of itself.

Since God is infinite and infinity is not conceivable by finite minds, God in his own nature cannot be an ideal that can be practically pursued by ordinary mortals. The ideality of God, his absoluteness, therefore stands in need of metaphors, metonyms, or some other translation into a human scale of intelligibility. Thus the ideal of the Christians is Christ, of the Buddhists, the Buddha. Bankim asserts his belief that no other religion has such a rich series of anthropomorphic ideals as Hinduism does—'they are detached even when ruling empires, they enquire into religious truth even when lifting the bow, wise even though kings, loving to everybody even though powerful'.[30] Particularly, in Hindu theory there is an ideal of such ideals which gives unity to this tradition theologically, philosophically and narratively. This is the figure of Kṛṣṇa.[31]

In support of this thesis Bankim advances a somewhat curious argument. A master narrative's ability to integrate a civilization does not depend on its actuality, *through* and in other happenings. The fundamental quality of the Oedipal stories does not require that Oedipus must be a historical figure. Indeed, one could say that to be an ideal, Oedipus, like the mode of production, had better not be an actual historical fact; for his task is not to be true himself, but to be true through others, to be true in every man, and he can exist in every human being without existing in the ordinary sense. Any ideal figure, on this logic, can do his work without being actual; indeed, it could be argued that he could accomplish his task better that way. For then the debate about the meaning of his life could not be displaced into an unnecessary controversy about whether somebody called Kṛṣṇa did ever exist or not; and whether the historical Kṛṣṇa pursued a life that is entirely free from all conceivable blame.[32] It could then be a more straightforward comparison and moral measuring of one ideal of humanity against another.

I think Bankim did not take that more obvious line of celebrating Kṛṣṇa as a pure ideal because he was caught in the colonial reprise. His celebration of Kṛṣṇa is in one respect entirely different from earlier theological disputes internal to Hindu or Indian religions. His celebration of Kṛṣṇa is not aimed against other, Indian or Hindu constructions of this ideality, but the ideality of Christ; and Christ was a *historical* figure. Bankim must therefore argue the historicity of Kṛṣṇa, not only his rational superiority as an ideal.

Bankim's strategy in this enterprise is simple and severely rationalistic. Events in the life of Kṛṣṇa are recorded in the *Mahābhārata* and nine out of the eighteen *Purānas*. Indeed, the extant versions of the *Mahābhārata* themselves point out, in a move of intertextuality common to Hindu religious texts, that the later parts of Kṛṣṇa's life are recited in the *Harivaṃśa* (which is known as the *Khilaparva* of the epic) and the *Bhāgavata*. Bankim notices, without appropriate wonder and without critical sympathy, a rather curious fact about these allegedly interconnected legends. A primary difficulty in reconstructing the life of Kṛṣṇa is the evident discrepancy between the accounts we get in the *Mahābhārata* and the *Purānas*. Bankim's analysis reveals a pattern of narrative construction in which later poets take advantage of the underdetermined segments of the earlier narrative and insert interludes which serve their doctrinal interests. In the *Mahābhārata* Kṛṣṇa appears as an already grown up person, a warrior with an apparently uneventful earlier life, at least, not significant enough to be recounted in great detail. Scraps of determinate features are available: his lineage, his living in Vṛndāvana, his undoubted prowess during his unreported adolescence. Later, the *Purāna* writers take advantage of the underdetermined adolescence of Kṛṣṇa to place the story of the *gopīs* and Vṛndāvana within this empty space of the narrative. As no one could be an adult without being an adolescent, and since Kṛṣṇa's adolescence was underdetermined in terms of facts, this was not, at least in a literal sense, objectionable.[33] The insertion of this sub-plot into the narrative of Kṛṣṇa was however immensely significant for later vaiṣṇava doctrine. This second, erotic narrative was to grow rapidly into an independent segment, and since this was laden with theological and doctrinal meaning, this made it possible to shift the whole philosophic centre of the Kṛṣṇa story in a new, emotivist direction consonant with *bhakti* movements. In course of time the Kṛṣṇa cult would come to embody an emotional religion of bhakti and eroticism. Gradually, this erotic tale about Kṛṣṇa and the gopīs undergoes constant elaboration. But *Mahābhārata*, Bankim argues, is unmistakably the first and the original stratum of the Kṛṣṇa story, a pre-condition for all that was to follow; other texts merely 'fill the gaps in the story structure of the *Mahābhārata*'. The *Mahābhārata* is therefore the original or primordial (*moulik*) and constitutes the fundamental and the earliest stratum.[34]

Fully presented, the argument about the authenticity of the

Mahābhārata runs as follows. The *Mahābhārata* story gives the fundamental and true account of Kṛṣṇa's life. The story is full of gaps and leaves large empty intervals, but is still true. If later writers fill in these gaps in the narration, there could be nothing wrong in principle, but, in fact, there is, because the kind of stories with which they fill in these gaps, tend to shift the philosophical balance of the story from his later life of practical action to his earlier life of abandon. These changes make it not only a much longer story, but a story of a different kind. Only formally is it the same Kṛṣṇa, in fact, he has been turned into a symbol of a very different doctrine of life.

Still, it could be said that the historicity we are talking about is the historicity of the accounts, not a historicity of Kṛṣṇa—what the accounts are accounts of. Bankim does not stop to consider this. Is the *Mahābhārata* really history? Obviously, this is a question where the answer would depend on what counts as history (*itihāsa*) in this case. The *Mahābhārata* possibly contains a great deal that is fantastic, impossible, plainly ahistorical; but that is not a decisive argument. First, there is a theoretical statement in the Hindu tradition about what is itihāsa.

> *dharmārthakāmamokṣānāmupadeśasamanvitam*
> *purāvṛttakathāyuktamitihasaṃ pracakṣate.*[35]

A subtle and evasive definition of 'history' which turns what rationalism regards as historicity into merely its minor clause in the definition: it is a story taken from chronicles of earlier times which gives instruction on *dharma*, *artha*, *kāma* and *mokṣa*. It is an interesting definition of what 'history' is, particularly in its frank avowal of the idea that 'history' only partially records the past, at least equally significantly it provides the present with justifications; but this is hardly compatible with the strongly rationalistic version of history which Bankim invokes, in which the claim of history to respect as 'positive' knowledge lies only in its quality of being true.

Bankim places great emphasis on the rationalist cognitive criteria of the historical enterprise. In ancient India, although people knew the art of writing, texts were generally disseminated and preserved through an oral tradition, and Bankim rather curiously thinks that modern rationalistic techniques of discovering inaccuracies in copies of written texts could be unproblematically applied to 'oral texts'.[36] Apparently, this phrase was used to mean that there were some texts which, though written, depended for their transmission on the oral tradition. Ordinary people would know about the *Rāmāyana* and *Mahābhārata* not by reading them, but by having them periodically retold, through a continuous and intriguing web of repetitions. It could be an interesting question to find out if all cultures allow such retelling of texts or whether some of them enforce stricter acquaintance with the original (e.g. the *Quran* and the *Testaments*). For Bankim, to find out the 'truth' of Kṛṣṇa is not to find an artistic truth in which one reading does not displace 'the truth' of other possible and very divergent readings, but a strictly cognitivistic truth of empirical facts which would dispel factual errors.[37] If what we are seeking is this cognitivistic historicity of the figure of Kṛṣṇa it follows that the text likely to be the most reliable would be one that is nearest to the life of the actual Kṛṣṇa in time. The *Mahābhārata* offers the most undistorted account of his life story. However, it has been a particular victim of the creativity of the oral tradition, and Bankim suggests that two types of stratifications could be found in the commonly available text—a historical and an artistic stratification—which, although using different criteria, reinforce each other. The original *Mahābhārata* has a fairly detailed *parvasangraha*, and whatever is significant for the story structure but is absent from this could safely be identified as a later interpolation. To critically separate the original stratum from the encroachment of later deposits and inauthentic interpolations thus becomes the crucial task.

Not surprisingly, the criteria that are used for selection are strictly positivistic; indeed, these are explicitly derived from ordinary scientific practice. 'All activities which constitute human life can be accomplished by depending on the logic of proof. We require stronger or weaker types of proof depending on the type of case we are going to judge . . . That is why for different types of subjects or occasions, different types of proofs have been recommended. For example, for law courts, the law of evidence, for science, inductive logic; similarly there is a distinctive logic of enquiry for the science of history'.[38] Bankim is entirely conscious that he is applying principles of proof from this external (western) tradition to understand the truth of an Indian epic. Contrast this, incidentally, with Tagore's defence of a non-cognitivist truth of a literary text, in fact, of the other epic, *Rāmāyana*:

> *sei satya jā racibe tumi*
> *ghate jā tā sab*
> *satya nahe kavi taba manobhumi*
> *rāmer janamasthān ayodhyār ceye satya jeno.*[39]

[Truth is what you create/write; whatever happens is not always true, poet, treat the land of your imagination as more real than Ayodhya, the land of Rāmā's birth.]

A non-cognitivistic theory of truth like Tagore's (or if you prefer a similar modern version, Gadamer's) would have gone in some such direction; but Bankim's strong cognitivism takes an entirely different course. On the application of such rigorous criteria rather a great deal of the Kṛṣṇa myth becomes historically suspect. Particularly because the cognitive controls Bankim specifies are lexically ordered: an episode, to be held true, must pass all his seven recommended tests. We are urged to reject whatever does not figure in the *parvasangrahādhyāya* and the *anukramanikā* and all mutually inconsistent episodes. Artistic creations of great authors usually display some strong stylistic signature, Bankim argues, and anything inconsistent with such peculiarities of style are interpolated. If anything is irrelevant to the story, and shows any of the earlier failings that is to be rejected as well.[40]

This formal apparatus of exegesis when applied to the *Mahābhārata* (which Bankim does not show us in detail) shows the epic to be unhomogeneous and stratified, and reveals three clear strata in its art. The first stratum, the original, was probably quite brief, confined simply to a recounting, without much luxuriating symbolism, of the life of Kṛṣṇa and the Pāṇḍavas. There is a stratum overlaid on the first with entirely distinct stylistic and artistic attributes. Some parts of the *Mahābhārata* are morally serious and uplifting, their depiction of human acts undistorted and natural, and of poetic quality of a very high order. Other parts lack this high tone, but are deeply connected with its underlying metaphysical doctrines, and therefore poetically unnatural, although certainly not without some poetic quality.[41] At this level poetic ingenuity is basically involved in showing the miraculous 'possible'.[42] Without the first layer the *Mahābhārata* loses its narrative coherence; if the second is taken out it does not lose anything vital except some decorative frills. The most significant difference between the two layers is shown by their treatment of Kṛṣṇa: on the first level, he never reveals or uses his divine omnipotence and accomplishes his tasks by consistently human means. In the second he is clearly marked and worshipped as an incarnation of Viṣṇu and is himself inclined to making this not exactly modest announcement rather often.[43] Curiously Bankim overlooks the obvious difficulty that on this view of stratification of the epic text in the entire Hindu tradition, the *Gītā*, his favourite text, would almost certainly be disqualified.[44]

The *Mahābhārata* also contains a third stratum: 'whoever ever composed something and felt pleased with his work simply squeezed it into the *Mahābhārata*',[45] an extraordinary proceeding by any standards, and hinting at a different theory of textuality. By the ordinary rules of reading and preserving texts, this would be a strange way of showing regard for it. But Bankim is not alone in taking this exegetic line; Rajsekhar Basu, in his influential Bengali abridgement of the *Mahābhārata*, offers a similar theory about *Kumbhīlakas*, an inverted *plagiarism*.[46] The *Mahābhārata* was a major source of instruction for common people and the Hindu tradition, according to Bankim, was given to systematic translation of the metaphysically abstruse into the metaphorically attractive; and the third layer of the epic is merely traditional Hinduism's department of public instruction, to be discounted in any serious analysis of its philosophic or aesthetic truth.

A consequence of this selection is the resolution not to accept miracles for we cannot justifiably believe in supernatural occurrences. 'We ought not to believe in supernatural events even though we seem to have direct evidence of these. We cannot believe in them even if we see them with our own eyes. For our senses can make mistakes; but contravention of a natural law is impossible. Only if you can prove to me that what I regarded as supernatural is something possible within natural laws shall I accept it'.[47] Most significantly, there is a final conundrum: 'If he is to accomplish his deeds through divine powers why did he assume human form in the first place? To destroy a few demons (*asura*) or common monsters is evidently not a difficult task for the omnipotent. If he has assumed human form he could not have intended to use his extraordinary powers'.[48] Thus we get a complex mix between a theory of the avatāra, a translation of the absolute into human scale, and an image of a Cartesian god.

In the first theoretical part of the *Kṛṣṇacaritra* there is some consideration of the last question, which shows explicit influence of European-style rational theology; 'Is it possible for God to exist on earth? Ordinary Hindus believe Kṛṣṇa was an avatāra of Viṣṇu; the educated consider this wholly unscientific, and Christian missionaries consider it laughable'.[49] In his response to these doubts Bankim divides the question into two: first, whether God could come down to earth; and second, if so, then was Kṛṣṇa an incarnation of God. Politically, this division is really a stroke of great force, for this utterly neutralizes part of the Christian opposition, as they must assent to the first possibility, as otherwise Jesus could not exist.[50] Bankim then undertakes to answer agnostic doubts of the scientific or philosophical kind about the second question.

Some maintain that God does not have attributes or qualities (*nirguṇa*), for to have qualities means to be finite. At the same time only a God with qualities or at least a God willing to assume qualities can have incarnations. 'I cannot conceive what a qualityless God would be, and therefore I cannot solve this difficulty. I know of course that a large number of learned and religious people regard God as being without qualities. I am not a learned man or a seer, but I suspect that these philosophers and mystics too, like me, are unable to conceive of a God without qualities, because man does not have a cognitive faculty by means of which he can possibly know an object without qualities. God may well be without qualities, but we cannot know such a God'.[51] 'Of course we can say (as opposed to conceive) that God is without qualities, and we can also construct a whole philosophical theory on this idea. But it simply does not follow that what we can utter is also something we can mentally conceive. It does not destroy our tongue to say "a square circle", but we cannot understand what it could possibly mean . . . If we say God is without qualities we lose the creator'. There are still others who hold that God is with qualities but without form (*nirākāra*), but this objection is easily answered. If God is really omnicompotent even though He is usually formless what prevents Him from acquiring form if He so desires? Another objection could be that if God is able to solve His difficulties by His powers why should He undertake the endless suffering of a human life? This forgets (a) that since the nature of God is different from our (sentient) natures, He must be beyond suffering and pleasure; and this sort of solicitousness about His possible suffering simply arises out of a mistranslation of His nature into our terms.[52] Particularly, because (b) what is a whole lifetime for man is an instant of His time, though, strictly speaking, if the first argument is right, the second is really redundant.[53] In fact, the necessity of God's incarnation does not lie in the minor irritations represented by a Kaṃsa and Śiśupāla, but in a larger generic necessity of destroying evil in general, *Dharmasaṇrakṣana* is not accomplished by just destroying a few chosen evil-doers. Dharma is the 'fullest development, completion, consistency, coherence, fruition of all our mental and physical faculties',[54] a point Bankim demonstrates in dialogue form in his *Dharmatattva*, which makes the whole series of arguments through a systematic counterpointing between European (Enlightenment) and Hindu definitions of common terms.[55] Such development of character cannot be attained simply by discourses made up of words. It requires a syntax of acts, a text of a concrete life, a chain of perfect actions. God's nature is irreducibly different from ours, therefore it requires incarnations of God to show what perfection of virtue means in human terms. It is God's way of circumventing the cognitive finiteness of the human mind.

Possibly some could object from a Cartesian angle that it is true that God exists, it is also true that He has created and governs everything, but it is irreverent to think of him as running the world 'in the same way as a coachman runs his coach'.[56] He has simply created some fixed laws for the existence of the world which are both immutable and sufficient for its continued existence. To believe that God would occasionally have to appear as a moral repairman of the universe is unseemly. Correction of its ills ought to be left to a sort of moral hidden hand. Even this argument does not appear decisive to Bankim. Although the laws of creation are immutable, it contains a law which requires constant change in a preferred direction; it is part of the laws of this world that the world would improve (*unnati*). Nothing in this world is in such a state that it could not be improved by God—a Hinduism with a clearly Spencerian flavour. 'If improvement is still possible how can we say that there is no possibility or necessity of divine intervention? Apart from creation, preservation, maintenance and destruction there is another *natural function*[57] in the world—improvement'?[58] And God seems to be very deeply involved in this new eternal process discovered (or begun?) in the second part of the nineteenth century. On this point,[59] as Partha Chatterjee suggests perceptively, we must look at the subtle translation of the functions of the Hindu

God in terms of rationalist and evolutionist thought of the nineteenth century. Bankim conceives of the irreplaceable function of the ancient incarnation of Viṣṇu to be energetic preservation of the laws of Cartesian rationalism, Darwinian evolution and Comtean sociology.[60]

Finally, the task of the avatāra is adequately specified: there are, it appears, two circles or levels of causality, one for God in His absolute and infinite self and another for men; and one quality that God shares with western philosophers of the Enlightenment is love of consistency. He tries to figurate dharma in His being in a human form; He shows what it is to live a good life. Bankim quotes a wonderfully playful passage from the *Viṣṇupurāna* on how God dissembles a human form—'The fact that He shot a great number of weapons against His enemies is part of His playful pretence in accordance with the human form. Why should otherwise there be such enormous exertion on the part of one who can create and destroy the universe by thought alone? As He is following the practices of men, He would make peace with the strong, fight those who are weak, use the techniques of equity, gift, division for justice, and (to make the performance credible) He would even flee on rare occasions. These infirmities of the human type were done by His own wish following the rules of human conduct'. To show what a good life means is literally to show it; not to suggest it through the abstract and symbols. He shows good life unmetaphorically by the directness and presence of His acts, through the non-abstract and, in a non-literate culture, universally understandable text of a life. His life, emphatically, is a text; because it fixes, objectifies, makes indisputable what to live a good life is (not what a good life means). He does it entirely consistently, within the terms He has set for Himself. It would be to stipulate a limit on this omnipotence to believe He cannot put limits like these on Himself. Since He has taken upon Himself the implicit resolve to stay within the limits of human causality including, if the *Viṣṇupurāna* is to be believed, an occasional flight to produce credence, He must remain consistent; for otherwise, this would simply mean deluding His potential followers, inducing them to take on irrational risks. Otherwise, would it be consistent with His infinite wisdom that He is incapable of understanding the virtue of the simple rule of consistency?

The Construction of Kṛṣṇa

What follows in the *Kṛṣṇacaritra*, after the procedural preliminaries are over, is a systematic reconstruction of Kṛṣṇa's figure in this double polemic, through the shifting of its centre of gravity from the affective and erotic to the rational. In Bankim's hands his reincarnation is so consistent and complete that he emerges as a more eligible candidate than Christ for the task of preserving order in a Cartesian universe.

It is essential to reconstruct correctly the terms of his question. Bankim's vindication of Kṛṣṇa is not aimed against other Hindu or Indian constructions of ideality, but the rational theology of Christianity, the ideality of Christ as constructed by rationalist Christian discourses of the nineteenth century. Christ was a historical figure. Bankim must, or at least he thinks he must, argue the historicity of Kṛṣṇa, not only his rational superiority as an ideal. What is interesting is not only the explicit debate, but also the silences: it is not only Christ and Buddha who are historical figures, but also Mohammad, who does not figure in Bankim's list. It is not always easy to read such silences. It is at least possible to argue that this shows where his crucial line is drawn; it is not between Hinduism and the other religions which would have made a debate with Islam obligatory but between Christianity and Indian religions. In any case he was familiar with the line of thought tried out by his contemporary Bhūdev Mukhopādhyāy that Indian Islam was different doctrinally and structurally from Middle-eastern Islam, and therefore, could be subsumed under a scheme of religious life crucially determined by Hindu influence. This in itself reflects a puzzling and generally unexplained problem; that Hindus saw Islam as a *political*, but not an *intellectual* threat. Their intellectual response to British and Christian doctrines is marked by a much deeper anxiety.

The polemic is a double one: it engages in a controversy against Christian theology, and more directly with European contempt for Hindu theology; but, at the same time, he believes that the argument could be won for vaiṣṇava Hinduism only if Kṛṣṇa's figure is seen in one particular way, not the way it had been interpreted traditionally. Bankim's argument therefore tries to refute westernism by refuting tradition. We must digress into a brief account of the interpretation of Kṛṣṇa in the vaiṣṇava tradition of Bengal in order to have a clear view of the doctrine he was trying to reject.

In the Gauḍīya vaiṣṇava tradition, around the mediating figure of Caitanya, a fundamental transformation takes place of the older warrior-rationalist figure of Kṛṣṇa, a man of action and a serious philosopher of praxis. Gradually, the rationalist emphasis is entirely subordinated to an emphasis on something that is an inextricable mixture of the erotic and the aesthetic. Eroticism is given a philosophical meaning and is seen simply as an ability to see happiness in the world, the availability of its beauty, the ability to see beautiful things in the everyday, and making the life of man beautiful by beautiful and kindly acts. Caitanya exhorts people to treat both the world that is available and their own lives as works of art. An aestheticization of life takes place through his interpretation which is intensely and complexly poetic. Caitanya is not merely able to see the poetry of the world, but also the poetry of words which evoke and celebrate it.[61] Caitanya, at the same time, interprets *premabhakti* in a way that makes it less and less dependent on intricacies of learned texts, arguing implicitly that the *bhakta* need not use intermediaries who can disentangle textual complexities, precisely because the glory of the text of Kṛṣṇa's life is in being a text that is so direct, simple and so eminently readable. Its beauty lies in the fact that it can be read without knowing how to read. Writing things in words, on this view, is more a hiding than a revealing; because special techniques are required to decode them. Socially, this tends to create a class of brokers between God and His devotees. God could not have wanted to write His message in such an élitist and inaccessible code. If He is *samadarśī* His code must be available to everyone, the most universal code of pleasure. The universality of sexuality does not divide human beings into minute unequal fractiles; it is the form of pleasure that is most universal, most available and also, interestingly, the most incommensurable. Sexuality and play (*līlā* or *vilāsa* are the words which indicate these two ideas) therefore provide the perfect metaphor for a religion of a non-élitist aestheticism.[62]

Naturally, this sexuality, literal or metaphorical, cannot flower without a suitable anthropomorphic figure, and Rādhā answers this need perfectly, gaining gradually, along with the transgressive tendencies of Caitanya's doctrine increasingly transgressive determinations. Caitanya is seen by his biographer as *Rādhābhāva* and *Kṛṣṇaswarūpa*; he has the physical and psychic characteristics of Rādhā, although in a deeper sense he is Kṛṣṇa's self. The *Caitanyacaritamṛta* puts this clearly in the fifth sloka of the first canto:

> *rādhākṛṣṇapranayavikṛtirhlādinī śaktirasmadekanatma*
> *navapi bhubi purā dehabhedam gatau tau*
> *caitanyākhyam prakatamadhuna taddvayam caikyamāptam*
> *rādhābhāvadyutisuvalitaṃ naumi Kṛṣṇaswarūpam.*[63]

The earlier alienation of Rādhā from the self of God is explained by an argument at once poetic and philosophical. And this contains a philosophic explanation of the necessity of Rādhā within the necessity of Caitanya:

> *srī rādhāyah pranayamahimā kīdṛśo vānayaivā*
> *swādyo yenādbhutamadhurimā kīdṛśo vā madiyaḥ.*[64]

Quite independently of Caitanya's own predilection for the erotic metaphor,[65] the Rādhā figure emerges out of a powerful tradition of erotic poetry; and Caitanya, for his own purposes, instead of condemning it for its sensuality, simply adds to its poetic and metaphysical power while advising the disciples that these are of a metaphorical, not literal status. In Gauḍīya vaiṣṇava theory, ironically Kṛṣṇa is gradually relegated to a distant eminence; and the whole metaphorical ground is gradually occupied by an increasingly resplendent Rādhā. The cult of eroticism undergoes a subtle internal transformation—the love, and its many states which are described become increasingly the states of love of a feminine figure, instead of a male. This is a significant and largely uninterpreted change; but for all its going unnoticed, it is quite unmistakable. One is tempted to claim that the classic tradition sees this transaction of love as being dominated by the male, the folk tradition by the female. Although that might be an untenable overstatement, in the *Bhāgavata*, and more proximately in the *Gītāgovinda* the male and female figures are equally active. An address like the famous passage spoken by Kṛṣṇa in the tenth *sarga* of the *Gītāgovinda*:

> *vadasi yadi kincidapi dantarucikaumudī*
> *harati daratimiramatighoram etc.*

becomes increasingly rare. In Vidyāpati, Caṇḍidās and other *padāvalī* poets Kṛṣṇa becomes literally a literary excuse, and disappears from the active verbal and descriptive circle. Now poets still describe happiness, sensuality, suffering and sorrow—but these are all Rādhā's states of being.[66]

It is particularly important to appreciate Bankim's intellectual daring in this undertaking, and we can understand the immensity of his task only if we see what is at stake. An entire artistic, or cultural, tradition hangs on this story, particularly in its popular form; and it will be a false simplification to believe that it is merely a folk tradition that is centred on this narrative. As the story assumed purity and refinement of form, and its peculiarly flowing and joyous structure, it became the centre of an increasingly large and unsubstitutible tradition in classical and semi-classical music, folklore, and a deposit of imagery and symbolism, which mediated the ordinary Hindu's approach to his world. What is interesting is Bankim's simultaneous delineation of an equally inexorable line of distinction between his view and that of the modernists who would look upon all this with contempt, or dump them as objects of anthropology. To try to get away from one's past by turning one's back to it is simply to misunderstand the nature of the historical reality of the present.

Bankim's actual reconstruction of the Kṛṣṇa figure—what he would have called restoration in the strict sense—follows a dual strategy. Fortunately, the two operations converge in their result. Kṛṣṇa's figure, Bankim thought, must be freed from all supernatural elements of Purānic lore. This would mean rejecting an enormous segment of the figure as it really existed as a cultural fact: His ability, in many different ways, to do impossible things.[67] Strictly speaking, these impossible accomplishments were of two types—materially impossible acts, and exaggerations. Exaggerations are now generally seen as an inextricable part of the popular tradition. Not surprisingly, Kṛṣṇa's powers were exaggerated in two spheres that mattered; in the field of war and combat, he can conquer anyone, and everyone put together, and in the field of love—in which again he can literally conquer anyone (including Rādhā, the symbol of erotic perfection), and everyone put together (in the *rāsalīlā*). On the night of the rāsalīlā he manages to cavort with no less than sixteen thousand gopīs, if the fluctuating estimates are to be brought to an average. The Kṛṣṇa myth is thus subject to one of the most constant and powerful of popular tendencies—it becomes a celebration of excess.

This doctrine of excess is turned by Bankim in a typically intellectualist/rationalist direction. Obviously, for serious intellectual argument in the post-Enlightenment nineteenth century, such excesses cannot be taken literally. Whatever socio-cultural needs these exaggerated exploits had fulfilled in earlier times, they had since disappeared. Rational reconstruction implied, consequently, that such exploits of Kṛṣṇa must be taken metaphorically. Bankim finds a different principle for restoring the integrity and integrality of the Kṛṣṇa image. We believe, then, that his life is described in two different registers, of the literal and of the metaphorical. Descriptions of what Kṛṣṇa does in the literal register do not go beyond what is humanly possible. Other 'acts' which find place in these narratives are metaphorical descriptions of his nature, falsely perceived as literal by poets of lesser ability. Not to believe in these exploits could not indicate lack of devotion for the Infinite, for the Infinite Himself undertook to restrict His acts within the circle of human causality when He undertook His avatāra. To emphasize the supernatural acts as marks of His infinite powers is both a moral and a cognitive error; a moral error because it is a mark of conceit to claim that we can understand the divine in His infinitude and a cognitive error because a large number is not infinity. Thus the different criteria of devotion and of rational belief coincide beautifully. Bankim's Kṛṣṇa can thus appropriate the Kṛṣṇa of popular joyous excess by deconstructing and retranslating him into a very different figure—from a figure of abandon to one of heroic action. He is transformed from a lovable popular figure of eroticism, excess, transgression, playfulness, a subject of both admiration and admonition, to a classic figure—calm, poised, rational, perfect, irreproachable. From the god of playful villagers and their folk festivals, a god who has to help them sort out small, everyday problems, He has been transformed into a god of a dependent nation who had to help them cross, nullify, reject and transcend in practice the historic indignity of subjection.

My attempt to construct a more or less rigorous argument out of Bankim's essay should not be misunderstood. It does not mean that he was in fact able to offer an entirely rigorous argument of the kind he was seeking. Nonetheless there are two reasons for us to systematize his views. First, there is enough of an argument of that sort to be reconstructed.

Secondly, it is precisely where the argument is not good, that we can understand him and his enterprise better, by simply shifting *our*question—from 'is the argument right?' to 'why does he think like this?'—an elementary interpretive principle, unfortunately not observed very often. Making sense of a text is to understand not only its truths but *also* its errors.

The way Bankim critically excises elements out of the popular Kṛṣṇa story is actually more complex than his rather severely simple criteria would suggest. A tradition is not infinitely malleable, and places limits on the permissible extent of internal change. Bankim takes the entire gamut of fantastic stories of the *Bhāgavata* through a double reduction—first, a naturalistic reduction in terms of a rationally acceptable occurrence; often this is aided by a second, subtler reduction that is linguistic and metaphorical. Interpretation dissolves an unnatural event into a metaphor or allegory. While the *Yamalārjunabhaṇga* is naturalistically read as an unspectacular display of credible strength, the killing of Pūtanā is reduced to Kṛṣṇa's surviving a children's disease, the taming of Kālīya into an allegory about the 'serpent' of time, the lifting of Govardhana into a metaphorical rejection of animism.

Understandably, the Kṛṣṇa myth which Bankim wished to attack and reconstruct is its erotic core—the story of Kṛṣṇa's life with the gopīs, particularly Rādhā. Here again Bankim uses a double argument for re-assurance. First, it is quite easy for Bankim to follow his theory of stratification of texts and critically excise the erotic parts of the myth. For the *Mahābhārata* has rare and fairly non-committal references to Kṛṣṇa's relations with the gopīs. 'This doctrine of the *vrajagopīs* is absent from the *Mahābhārata*; in the *Viṣṇupurāna* it exists in a virtuous form (*pavitrabhāve āche*); in the *Harivaṃsa* for the first time some amount of sensuality enters into it; the *Bhāgavata* expands on the element of *ādirasa* (sensuality); finally, there is a veritable deluge in the *Brahmavaivartapurāṇa*'.[68]

Three elements of the gopī story are reduced; by reduced I mean either entirely excised as an interpolation unworthy of the text or translated into a different register from literality—i.e., either into a metaphor (*rūpaka*) or some other form of linguistic reduction. Whether it is seen as a metaphor or otherwise, these are seen as statements which do not mean what they say, at least literally. I consider this reductionist interpretation of religious beliefs a distinctly modern phenomenon; it is an attitude present only in people who, like Bankim, have accepted a disenchanted view of the world constructed by science.[69] Since they find a literal reading of these beliefs implausible, they systematically deflect religious belief into two other possible registers: a register of ethics and a register of aesthetics, which are both, on the modern view, sufficiently independent of scientific language to have independent criteria of acceptability of their own. I think there is a serious anachronism in the view, often met in modern times, that ancient people, including makers of ancient religious and philosophic doctrines, thought of religious ideas as metaphorical in this strong sense. Seeing religion as primarily metaphorical is a product of acceptance of a certain material primacy of scientific discourse; it is the mark that the presence of scientific discourse leaves on religion. It is hard to believe that people of earlier times had such metaphorical ideas about religion, because they lived inside a discourse unthreatened by scientific theories. What is attractive about Bankim's treatment is that he admits candidly that prior acceptance of the validity of natural science forces him to interpret religious themes metaphorically.

The first element to be reinterpreted is the episode of the erotic līlā between Kṛṣṇa and the gopīs in the descriptions of *rāsa*; the second, the uncomfortable story of the stolen clothes, and finally, the absent centre of Bankim's reinterpretation, the figure of Rādhā, the foremost of the gopīs, the irreplaceable node of vaiṣṇava aestheticism. One of the major factors at play in all this is of course a concept of literary taste that is partly of western origin, in which Bankim assents to the babu's unquestioning acceptance of Victorian marital ethics, although in other places, as in his humorous masterpiece, *Kamalākānta*, he shows up its double standards, the tenuousness and uncertainty of its ceremonious distinction between public and private, and a hypocritical consigning of sexuality to the private domain. In *Sāmya* as well he has intensely indignant passages on the iniquitous consequences of this morality for Bengali women. Despite these and his occasional defence of the uninhibited literary styles of traditional poetry (as in Iswar Gupta) against babu Victorianism, the Kṛṣṇa myth is taken through a typically babu criterion on erotic taste. Bankim does not deny the poetic or aesthetic excellence of this myth, 'but we cannot pause here to taste poetic excellence, we are engaged in enquiring into far more serious theoretical matters'. It could be suggested that these elements of eroticism and uninhibition filtered into the myth because of the gradual mixture of transgressive stories of desire from folk tales into the highly classical image of Kṛṣṇa.[70] It is also historically possible, as I have suggested, that the actual social transgressiveness of the vaiṣṇava tradition emphasized the transgressiveness of the Rādhā story, although this no doubt helped them to communicate through its metaphor of intentness some purely metaphysical ideas about the relation between the devotee and his god.[71] Bankim is not concerned with the history of the myth but with its content. He accepts that the behaviour of Kṛṣṇa with the gopīs, if taken literally, is 'against the canons of modern taste'.[72] Perhaps we should not misunderstand this as the mark of a timid Victorianism, because Bankim's handling of love in his novels has a materiality of desire quite different from the unconvincingly ethereal treatment of love in Tagore and lesser Bengali novels, where persons are usually incapable of experiencing anything beyond purely conceptual passion for their beloved. Indeed, in

Tagore, despite his obvious sense of physical beauty, passages of direct physical description are rare; occasions for physical contact in his novels would be statistically insignificant. However, in this case, Bankim accepts the bhadralok equalization of the common obscenity of *battala* literature and the material eroticism of religious myths, and the task is obviously to exorcize the element which offends the taste of a colonial Victorianism.

Bankim decides to treat these motifs as a complex and interconnected structure of allegories, particularly material allegories of indistinguishability and surrender—which are beautifully expressed through an erotic symbolism not only in vaiṣṇavite Hinduism, but many other religions.[73] If he is able to get his readers to translate both rāsa and *rati* as mere play, on which he quotes the authority of Srīdharaswāmī, the impact of the excesses is considerably reduced, translating the wildly orgiastic into the excessively playful. There is an undergirding philosophical argument here. Of the four *vṛttis* or 'faculties'[74] (Bankim himself suggests this translation in the *Dharmatattva*)—after the physical, cognitive, practical, the final one is aesthetic (*cittaranjinī*) and the rāsalīlā is then a metaphorical assertion of the idea that in God the ability to get pleasure from the aesthetic availability of the world is infinite. To be one with God, conversely, individuals must cultivate their aesthetic faculties, no less than cognitive and moral ones; to be able to participate in the joyousness of existence (ānanda) and aesthetic enjoyment is a way of being with God. Accordingly, the rāsa takes place in a flowering resplendent nature, symbolically, on an autumnal full moon night, and this could be read as nature's manifestation of its union with the Eternal.

In the case of *vastraharana*, Bankim takes an entirely defensive posture, and regards this as contrary to canons of refined taste. This is an odd idea, precisely because so much of the current idea of taste is drawn from the European tradition, and the babu culture could not have been unacquainted with the ubiquity of the nude in European painting. Perhaps he is defensive about the element of sexual fantasy often characterizing depictions of the episode, particularly in the *Bhāgavata*.[75] Philosophically, this simply signifies, through an unnecessarily lurid metaphor, the conquest of shame, the last vestige of self-consciousness. Only those who have truly conquered even the last, even the most winning vestiges of self-awareness—like shame in women—can aspire to know God. Although Jayadeva does not describe this episode, Bankim is particularly severe on Jayadeva: 'what is still a deep religious doctrine in the *Bhāgavata* has become a festival of sensuality in the hands of Jayadeva Goswami',[76] and the way the Purānic authors depict the *viraha* of the gopīs reminds him, I think quite rightly, of the entirely material *viraha* in Kālidāsa's secular poetry.[77]

The Figure of Rādhā

Bankim's final reinterpretative effort has to take on the most powerful of all such figures—Rādhā, obviously because without Rādhā, or with a linguistically reduced Rādhā, the figure of Kṛṣṇa undergoes a completely antierotic transformation. Rādhā is the final metaphor of vaiṣṇava religion, a figuration of *bhakti*.[78] In vaiṣṇava doctrine Rādhā is not only the symbol of devotion, Rādhā is the alienated self of Kṛṣṇa, his *śakti*, his powers in abstraction. Kṛṣṇa, to extend the metaphor, is apparently subject to the paradoxical limitation of all selves. As understanding, even seeing is a communicative act, the self can see itself only if it sets up a relation with itself of an other, the way the physical body needs a mirror. Rādhā in that sense is the mirror of his self. The *Caitanyacaritamṛta* says that it is because of the need for self-knowledge that '*dehabhedam gatau tau*', they undergo a split into two forms: for the doctrinal purposes of the *Caitanyacaritamṛta* of course this is particularly helpful, because what it wishes to say about Caitanya is wonderfully enhanced by this theory. In Caitanya, Kṛṣṇa reincarnates himself to understand exactly what Rādhā feels about him. He is the sublation of their duality. That is why Caitanya must be *rādhābhāvadyutisuvalita kṛṣṇaswarūpa*. Bankim is less centrally interested in Rādhā; he is more concerned at what happens to Kṛṣṇa when Rādhā is around; he is seen as *dhīralalita* whose characteristics are disastrously recounted as follows:

> *vidagdho navatārunyaḥ parihāsaviśāradaḥ*
> *niścinto dhīralalitaḥ syāt prāyaḥ preyasīvaśaḥ.*[79]

For Bankim this is a sure recipe for disaster, as this Kṛṣṇa is not too different from the babus produced by Hindu College. The role of the avatāra is to show a complete form of humanity, and this will mean that the ideal will dissolve into a life of uncontrolled sensuality, inactivity and purposelessness. Thus, although the traditional vaiṣṇava argument that Rādhā is the alienated self of God created in order to understand his own nature is philosophically unexceptionable, Bankim prefers a rather different route. This is a route of linguistic reduction.

The tale of Rādhā and Kṛṣṇa in vaiṣṇava theory must indicate something more about the nature of accident and necessity. Otherwise it is difficult to see the logic behind a construction in which the lord of all creation, who creates not only the world's determinations but also the necessities, was not able to get round such accidents of deprivation—as being born younger than his beloved, or not being lawfully entitled to her hand in marriage. This is an extraordinary failure on the part of the Almighty. Evidently, the lord of all created things could have managed His birth a year or two earlier than He actually did, or been born into a family from which He could have courted his alienated self with greater legitimacy, or at least prevented her vague but regrettable marriage to an unspectacular cowherd. Curiously, He did neither of these things, and decided, unaccountably to undergo the sufferings of forbidden love.

If His task is to present the world with an order, it is hardly the best way to begin by transgressing one of its most fundamental moral norms.[80] Bankim seems particularly bothered by this problem about the *Bhāgavata* story, and consequently tries to excise this part from the biography of Kṛṣṇa. This is obviously on the view that the lawgiver cannot be a petty lawbreaker without considerable strain on the internal consistency of this moral order. As time passes, assisted by the non-written quality of the tradition which does not lay down clear narrative boundaries, the figure of Rādhā, like most similar figures becomes more complex and less coherent.[81]

Initially, there is reason to believe that she is seen entirely as a metaphor, as a figuration of bhakti; she mystically disappears with the other gopīs after her mention in Kṛṣṇa's *pougandalīlā*. They figure in the story massively just this once, and then simply and irreversibly drop out. Though apparently remarkable, this is also easily explained. If we insist unreasonably on a realistic everyday reading of the story, the gopīs, one must presume, went back to their ordinary, quotidian course of life after the period, or the season, or the night, touched by divinity on the occasion of the rāsalīlā. Rādhā could hardly be consigned to such unreported anonymity, after the conferment on her of the more abstract and therefore purer quality as a figure. Thus, her suddenly but decisively dropping out of the narrative structure of the myth does not do much good to the consistency of its internal form.

Some aspects of the relation between Rādhā and the other gopīs from whom she gets gradually differentiated, are ambiguous and still raise problems of consistency. Initially she is merely the first among the gopīs. This will simply indicate that gopīs collectively are the symbols for devotees, who gradually through surrender of their selfhood, become real devotees of God. They represent collectively the *process* of acquiring devotion, achieving indistinguishability with God, she is the final form of this process, the rarest and therefore mortally unattainable; by accentuating this process of shedding the self she becomes identical with God, becomes in that sense His self. This however clearly raises a difficulty; for there is a crucial difference between evolving towards God from the side of the devotees and being on the side of divinity itself, representing its alienation into an other in order to understand its own nature.

A metaphor does not need material or ontological features necessary for dramatic or literary belief, and Rādhā consequently is more ethereal than ordinary fictional characters. She shares with some fictional entities the attribute of not having the basic ontological determinations mentioned. These missing determinations are not trivial ones like shoe-size; her birth, her own family (not her husband's family which is monotonously emphasized) are usually deeply and uncomfortably vague; as is her own peculiar fading away after the story of *braja*. She is not mentioned in a most radical way: she does not die, does not undergo any of the experiences reserved for common mortals. She is a metaphor; after the resplendent sexuality of her youth, it is only someone like her who can so easily, simply, finally cease to exist. That is her privilege as a metaphor. For the whole structure of Hindu religion, of traditional Indian artistic sensibility would collapse if she is allowed to be anything except her radiant self, if she is allowed to age. Even Kṛṣṇa himself, although not shown exactly rheumatic by age, is understood to age, and to be under the considerable strains of statecraft at the time of what must be regarded as Indian feudalism in its particularly bloody beginnings. Only Rādhā does not figure in the world in this way, and is miraculously free of its limitations.

Some later vaiṣṇava authors were seemingly dissatisfied by her sudden disappearance, and brought her perilously close to mortal sorrows. They invest her with ontological burdens of existing in the material sense. A real woman, unlike a metaphor, must have a before and a beyond; although she still lives within the terms of art, of fiction, she tends to return to Kṛṣṇa's story at various points, and is made thus to have a linear life. But because of the loving devotion of the poets she is exempted form the indignity of an actual old age. It would be inconceivable, indeed a disaster for the entire vaiṣṇava consciousness to encounter an old Rādhā—without her beauty, her youthful sexuality, her resplendence. In later vaiṣṇava poetic theology, however, there are some beautiful ślokas in which a new kind of wistfulness is created through the experiences of her later self. When she meets Kṛṣṇa after the war of Kurukṣetra she yearns in a wonderful śloka for her first union with Kṛṣṇa, against the necessary linearity of time:

yaḥ kaumāraharaḥ sa eva hi varastā eva caitrakṣapāste chonmīlitā
mālatīsurabhayaḥ prauḍaḥ kadambānilāḥ
sā caivāsmi tathāpi tatra suratavyāpāralīlāvidhau
revārodhasivetasītarutale cetaḥ samutkanthyate.[82]

Although sad and beautiful in a different way, these are yearnings which only persons with ordinary empirical determinations can have; metaphors cannot really have such sorrows.

Historically both the Rādhā and Kṛṣṇa figures get gradually transformed until they become unrecognizable incarnations of their earlier selves. In the course of this transformation Kṛṣṇa seems to suffer more than Rādhā, for although she is gradually interpreted towards a figure of exaggerated though picturesque weakness, Kṛṣṇa declines into complete inarticulateness. Literally, the later Kṛṣṇa hardly ever speaks; he is of course spoken about, but he becomes less and less a source and a creator of discourse. In the *Gītā*, he is at the height of his discursive self—indeed, in the *Mahābhārata* he does hardly anything but speak, and even when he has an impact on the narrative, it is essentially through discourses of some form. Gradually, this discursively sonorous Kṛṣṇa falls silent. Interestingly, in the *Gītāgovinda* he is not at all silent, and shares the discourse of love and sexuality with Rādhā, although one could say that there is a basic asymmetry in their speech. The speech of Rādhā is obviously a speech of authenticity. But on the evidence of the poem itself, in terms of the events which are purely internal to the *kavya* (i.e., if all that is reported of him is true), his speech must be false. I do not deny that through gaudīya vaiṣṇava theology we might be able to restore to his speech some metaphorical and metaphysical truth; but that itself proves my point that their speech in *Gītāgovinda* is asymmetric. What is important is that the later Kṛṣṇa is reduced to discursive nullity, little better than an excuse: his only artistic *raison d'etre* seems to be that he is the remote but necessary object towards which Rādhā's passion, laments and her increasingly tragic sense of the world is directed. The great stream of pathos flows towards him, making him in a peculiar sense both essential and inessential. However, the basic artistic fact seems undeniable. This is Kṛṣṇa's reduction to silence, Kṛṣṇa's increasing distance from this luxuriance of sorrow in this tragic discourse, perhaps because it is considered inappropriate to depict the lord of all creation unbecomingly broken by separation, and in an act of self-abasement.

In some ways the change in the figure of Rādhā is more paradoxical. She was absent from the *Mahābhārata* story; but by the time of the *Gītagovinda* she has already become its central figure. Yet it would be seriously misleading to suggest that since in a sense she remains the central figure of the Kṛṣṇa story from Jayadeva to Tagore's *Bhānusiṇher Padāvalī*, she remains the same Rādhā. Her transformation is no less drastic or astonishing. When she is still narratively indistinct in the early *Purānas*, the narrators add spice to their stories by speaking of a particular gopī who is characterized as *darpitā*, too arrogant about her gift of extraordinary beauty, who does not stop from scolding, rebuking, or bullying God Himself, because of the omnipotence of her gift. She is thus portrayed with a mixture of indulgence and rebuke: for the way she treats the lord of all creation is astonishingly immodest and wrong, but at the same time, in some subtle way, interesting, appropriate and beautiful. As her figure blooms into the later Rādhā it is these aspects of her which are emphasized. She is seen as *dṛptā*, self-respecting, assertive, often, with good reason, abusive towards Kṛṣṇa, *māninī*, *kalahāntaritā*. In these early texts she is hardly ever without strong self-respect, never representing an overpowering, though picturesque weakness. After all, one must not forget she is the self of God and theologically no less imposing than the Lord Himself. Indeed, she is purer in a sense in which He is not, in the way in which the quality of redness must be considered purer than an object that is red.

Gradually, in the Bengali vaiṣṇava tradition, her nature changes in a strange manner. The change is so drastic that in the narrative the *Vṛndāvana* episode is gradually overshadowed by the part designated as *māthur*;[83] it is a harder, deeper, more tragic separation, which bears a prefiguration of an eternal and unending alienation, unrelieved by realistic hopes of a reunion. In *Vṛndāvana*, on the contrary, separations were simply preludes to a deeper reunion, much like what happened to Kālidāsa's *yakṣas*—

ānandottham nayanasalialam yatra nānyairnimittaiḥ
nānyastāpaḥ kusumaśarajādiṣtasaṇyoga sādhyat
nāpyanyasmād praṇayakalahādviprayogopapattiḥ
vitteśānām na ca khalu vayoḥ yauvanādanyadasti.[84]

Later, the figure of Rādhā, from a metaphor of the joy of life, becomes a figure for its sorrows, its perpetual longing for an unattainable eternally elusive happiness. By the time of the later vaiṣṇava poets like Gnānadāsa we find her a one-dimensional wailing figure, perpetually on the verge of separation and despair. When Tagore recreates the padāvalī form in our own times, she has practically merged into the tearful helplessness of middle class femininity; she is shown as

ākula rādhā rijha ati jarajara
jharai nayanadau anukhana jharajhara[85]

whose only recourse now is death.[86] In this poetic incarnation she is more like the tyrannized Bengali middle-class widow rather than the forceful, proud, aggressive and above all joyous heroine who celebrates life through her resplendent sexuality, irrespective of social propriety, which confers a radiance on everything it touches—herself, Kṛṣṇa, nature, and the lives of distant audiences who can only hear that astonishing story retold by inadequate bards.

Despite monopolizing the discourse internal to the story, Rādhā goes into a decline as an artistic and metaphysical figure becoming a one-dimensional, weak, if poetic, bore. Actually she comes to merge into a different conception of femininity, which through the babu mediation, can mix without contradiction or resistance into the Victorian conception of the decorative feminine, which simply adores, celebrates and mythicizes women's vulnerability and dependence. The only excellence she retains is not in enjoying life, or rather, wresting her share of life's raptures, but in complaining, verbally embroidering on her suffering, emphasizing and developing a cult of her helplessness, her lack of subjectivity and her status as the ultimate victim. The damage is much wider and deeper, because with herself she carries the entire class of women into it. She can only get things for herself by invoking pity, exaggerating her dependence on other people's acts. From the central figure of the autonomous feminine, she had turned into an ideal of the cult of heteronomy. From a figure of incomparable joy, she has become the symbol of incomparable sorrow, from fulfilment to a waste of womanhood. Bankim's rejection of this evolution of a wan and martyred womanhood is clear; in his novels he clearly tries to reassert the other, earlier type—sensual, positive, with a forceful if occasionally frustrated joy of life. His heroines are in this respect closer to the earlier Rādhā, or rather the features of Rādhā are distributed among his female characters. Also the *Sākta* conception of the feminine is brought into the primarily vaiṣṇava iconography in a powerful combination in the underground temple in *Ānandamath*.[87]

The later Rādhā is clearly uncongenial to Bankim's theoretical intentions. He chooses two different routes of argument about Rādhā, and through her, about the more general question of what femininity symbolizes. One part of his argument is provided by the creation of images of an autonomous femininity in his fictional characters. The narrower task of dealing with the figure of Rādhā is accomplished through a technique of reduction.

Thus, Rādhā is first shown according to traditions of vaiṣṇava theology, as being Kṛṣṇa's alienated self, or His power, and therefore consubstantial with Him. Rādhā is a word coined out of the verb *rādh*, connoting worship in the form of a complete merger and surrender to God. Rādhā is therefore not an individual, not even metaphorically, but the name of a state of mind. Similarly, *braja* can mean the world; 'the town which is now called Vṛndāvana is manufactured by the vaiṣṇava Brahmins of Bengal'.[88] Bankim regards the seriousness of his reconstruction unrepentantly: for 'the task of the critic is twofold—destruction of superstition and the organisation of truth'.[89]

A number of ideological processes were combined in the complex transformation of the warrior and rationalist Kṛṣṇa into the erotic god of the eleventh–twelfth century. First, the earlier classical mould of character was broken, and a great number of pettier anecdotes brought into the story. Secondly, a strongly erotic element was introduced, which made the story more engaging, popular, but reduced its earlier tone of grandeur and solemnity. Third, since the erotic form requires an other, a sexual counterpart, this brought in and enhanced the character of Rādhā. Besides, the eroticized narrative mixed two dissimilar things: a metaphysical theory about devotional merger and the imagination about the pleasures and risks of illegitimate love.

By excising the figure of Rādhā through linguistic reduction, Bankim seems able to reverse or restructure this whole historical process. His work should be seen in any case as an explicit rebellion against current tradition in the name of a more distant and doubtful one. Each one of these successive historical changes is sought to be reversed by Bankim's argument. By calling Rādhā a mere idea, the attention of religious thought is brought back to Kṛṣṇa himself. Secondly, Bankim systematically disentangles eroticism of the folk from metaphysical arguments of self-estrangement. This still leaves a Kṛṣṇa who 'does a whole lot of things which are aKṛṣṇa', in Tagore's phrase, acts which are unKṛṣṇalike or unworthy of Kṛṣṇa's character. We are advised in such cases to be indulgent towards them only as *kāvya*, poetic license, not serious religious doctrine. We must, if we accept Bankim's theory, excise these folk elements from the story and return to the first, deliberately classical form.

Although Bankim does not say it in so many words, there seems to be an underlying suggestion that a people construct their god in their own image, and deserve in this sense the divinity they get. Lazy, purposeless, inactive, sensualist middle-class Bengalis, not unnaturally, worship a god who does nothing except chase women. This answers the question about the possible purpose of the classicization of the Kṛṣṇa figure. Substitution of the universality of eros by the other universality of reason, the painstaking excision of folk themes, its associations of sensuality and playfulness, leaves us a severely classical figure. It is, first, a figure of sufficient seriousness to contend with others like Christ, Kṛṣṇa's most serious competitor. It is this figure which can serve as a symbol of historic action, of practice. Practice of *swadharma* could create preconditions for the strength of the nation. But Bankim requires a recondite and exceptionally indirect argument to arrive at this conclusion.[90] Self-preservation is the ultimate and indefeasible right of all living beings, and therefore presumably, also of Indians. Defence of one's family and community is a simple extension of the basic right of self-defence. A further extension of this right could cover one's nation. Dharma is not religion; it is the right to be oneself, to express what one most fundamentally is. One cannot do that without defending and preserving the self. If to defend oneself, and one's legitimately extended selves is a part of dharma, it could also be part of one's dharma to organize and improve the strength of the nation to which one belongs. People must know what their dharma is, but this should not be confused with acquiring formal education; for a person who is educated is not necessarily knowledgeable, nor is the reverse of this true.[91] Education in Bankim's society is organized on essentially modern European principles which push to their extreme rules of division of labour and specialization. Educated men are consequently excellent half-men, or brilliant fractions, but less than complete human individuals. Development of a human personality means cultivation of all the four *vṛttis*, something that, on this showing, can never be accomplished by acquiring a western education. A truly religious person is one who is able to live a complete life which is the same as living a life modelled on the example of God.

Thus, suddenly and surprisingly, what could initially appear as a learned, occasionally abstruse dissertation on Hindu theology, takes a sharp curve towards a theory of contemporary praxis, a political theory in a particularly transparent code. This way the underlying link between this sustained reconstruction of the Kṛṣṇa figure and the practical energy of the vaiṣṇava warriors of his later novels becomes apparent. These people are surprisingly aggressive and violent for standard vaiṣṇavas, and they all incongruously march into violent battles singing vaiṣṇava chants now turned into slogans, which would ordinarily have only the quietest ring. This theology was actually a political field and was the subtle point and preparation of a great theatre of imaginary praxis.[92] Throughout this gospel of praxis—left vague, ambiguous, opportunistically interpretable, but unmistakable nonetheless—there is a latent interpretation of the *Gītā* and the life of Kṛṣṇa as essentially just and practical;[93] and there is of course the customary strategy of sharp but unsustained critical humour about the English and their accomplice, the babu. Here the contestation is not about reform, or legislation of a new plan for society; here the terrain is the more crucial and subliminal one of mythology.

A Myth of Praxis

What is the consequence of this classicization? What is supposed to happen through this long exegetic exercise? Formal definition of classicization is notoriously difficult. But the difficulty of definition does not nullify the genre. Classicization increases the distance of the object from the giver, from the way it is generally encountered in experience, the ordinary, the familiar. It is often by its unattainability that it becomes an ordering concept. The purpose of the ideality of God, it could be argued, lies precisely in the unattainability of His perfection. Excessive availability would defeat the purpose of having an ideal figure.

Bankim's own position is the somewhat curious mixture of an overt theology that asserts *Kṛṣṇastu bhagavān swayaṃ* and a Feuerbachian idea that God is a constructed object, an ordering concept, His life story a master narrative that a civilization constantly recites to itself because it wishes to believe certain things about what it itself is.[94] It follows then that if this narrative is reconstructed, if we tell ourselves a different story about what we are, we shall eventually be different.

Thus the narrative of Kṛṣṇa has to be reconstructed not for settling an antiquarian point about the past, but actually to strive in a particular manner about the future. The fault with its present form is its eroticism, its lack of rationalism, the fact that it encourages an attitude about the world that is contemplative and not practical. There is also a point of contest between the folk and the classical forms of Indian heritage. Folk forms bring the ideal too close to the quotidian, to the imperfection and untidiness of the everyday. The ideal of ideals of the Hindu civilization must be rescued from this ordinariness, from the entanglements of eroticism, and given purer intellectual outlines. That is why Kṛṣṇa must suffer a separation of sorts; he must renounce a material Rādhā in order to be united with a conceptual one.

At the heart of this reconstruction of the Kṛṣṇa figure, this richly complex mixture of theological doctrine, history and hidden politics, lay the great translation problem faced by the colonial intelligentsia. At a subtler and deeper level, there is a connection between the humorous bewilderments of Kamalākānta, his conundrums and this serious enquiry into the master narrative of Indian culture. In entirely different, occasionally quite antithetic ways, in utterly divergent genres, they wonder about, confront, and try to find solutions to the same cultural puzzlement. They are not two answers to the same question, but two aspects of the same answer. I do not consider Bankim's construction of Kṛṣṇa 'orientalist' as Partha Chatterjee has suggested.[95] That seems to me an interesting and fertile exaggeration. Fertile because, although it seems to me false in its extreme form, a serious criticism of Chatterjee's position helps us raise questions about Bankim, and more generally about nationalist thought, which would have remained unasked otherwise, and these questions are more complex and quite different from the usual certainties about Indian nationalism.

Orientalism, no doubt, was perfectly known to Bankim, because it was endemic in the Bengali culture of his times. But his rejection of orientalist conclusions seems to turn on a fundamentally non-orientalist distinction between rationalism and rationality. Rationalism is a western doctrine historically bound to the specific trajectory of the Enlightenment, to the contingencies of European history. It is part of the discourse of rationalism, its delusion and ideology, to see its own discourse as transcendental, as being true at all times, a discourse in which the origin is a mere accident, which does not, as in other cases, determine its inner structure. It is part of the delusions of western rationalism to forget its own historicity. Bankim questions this absent-minded, self-evident equation between human rationality and its contingent form in European rationalist theory, an equation accepted by the ordinary babu. Bankim appears to construe what he is doing here as a rational rather than a rationalist construction. Rationality does not consist in accepting as infallible the historically contingent, fallible, ethnocentric conclusions of a certain type of western philosophical theory out of which the ideology of colonialism arose. Macaulay's essay, he remarks in *Kamalākānta*, is indeed a work of fiction.[96] Truth does not lie, on the other hand, in asserting that not only rationalism, but rationality too is an alien product, that ancient Hindu civilization can defy rational criteria, that rational procedures of thought and practice are unnecessary and inapplicable to Indian culture. A rational solution to the problems of colonial existence was to produce a double critique—to take one's culture through an unmerciful reconstruction of this sort, and simultaneously by using tools of rational discourse against the false and exaggerated claims of European rationalist colonial ideology, by discovering and inventing rational ways within one's own tradition.

For Bankim this results in a deep intellectual loneliness. This figure of a rationalistic warrior-Kṛṣṇa bereft of Rādhā actually showed and confirmed his peculiar isolation in his age. In his dialogue, *Dharmatattva*, the disciple expresses this anxiety, and warns his preceptor that 'Hindus are not going to accept your Hinduism'.[97] Consequently, although in different ways, the theologian of *Kṛṣṇacaritra* and the opium-eater of *Kamalākānta* are equally marginal voices in nineteenth-century Bengal. Like several others, Rāmakṛṣṇa, the Brāhmos, the traditional conservative Hindus, Bankim too had to look for some implements for making arguments general, making them part of a general consciousness of society. In most other cases of reconstructed Hinduism the weapon of doing this was a movement, a sect, an organization of some sort. In this case alone, these implements are actually spurned. He can afford to do this because his 'movement' is the Bengali language itself. If he lacks propagators of his ideas among the members of Bengali bhadralok society, he has a permanent army of actors propounding his theory in the terrain of the babu's imagination through his novels. Although he did not structure the babu's daily social reality, he had permanently structured a part of his dreams, his literature.

I have suggested earlier that Bankim took a sort of personal revenge for the collective tragedy of his people by his humour, his double irony, by mocking at a self which is much wider than his self. In *Kṛṣṇacaritra* there is an attempt at a further step, to transcend this helplessness, to atone for this collective collapse in front of the practical syllogism of politics. He points to the hidden resources of the civilization, to symbols that were neglected, despised or falsely aestheticized, symbols of praxis, of avenging, of creation of self-respect for his collectivity. Through this myth there is a suggestion as to how to transcend the pervading sense of frustration.

History can never be free from a mixture, an undecided tension between what happened and what ought to have happened. Invented history is a way of mythicizing the collective self, extending it in all kinds of rationally impermissible ways, giving the present a worthier past than it actually had. Dramatization and mythicization are constant parts of men's lives; they are falsehoods by means of which people often try to cope with the overwhelming nature of reality. Thus Bankim's *santāns* are seen constantly stretching their forces mythically, mythicizing themselves because they wish to believe something which they know they cannot rationally believe about themselves. They turn themselves into myth in two ways. In the thick of battle, when the fortunes of war seem to turn against them, when they look behind they see not only their fellow soldiers of the day, of this particular battle. Stretching behind them into an infinite series in the endless expanse of time they must 'see' the ranks of all similar soldiers of history, finally transfigured by the subtle presence of Kṛṣṇa himself. For Kṛṣṇa after all is the reassuring sign of a fundamental order, the ultimate indefeasibility of justice. Even if the soldiers fail, they are already enacting the paradigm of martyrdom, in Victor Turner's phrase,[98] so that even in their failure they cannot absolutely fail, so that even their failure is a success of a kind. Actions of the martyr have a prior guarantee of success—either the triumph of victory or the triumph of death. This is rendered possible by a second element of myth: they look upon the present battle as an incident in a long and essentially unfinished process in which today's fighters may well lose, but someone else some day must win the war for them.

The idea of Kṛṣṇa himself being present is not facetious or a delirious illusion in the panic of the battle. The Kṛṣṇa myth provides precisely for a device of such generative recall. Does the *Gītā* not equate Kṛṣṇa with the principle of retribution, of justice, as a decisive answer to nameless injustices in the future? On the field of battle he spoke to Arjuna of his millennial anger:

> *paritrānāya sādhūnām vināśāya ca duṣkṛtām*
> *dharmasaṃsthāpanārthāya sambhavāmi yuge yuge*[99]

Yet, on a strictly rational reading there are problems with this curious reassurance because evil is everyday and the Lord's revenge is merely epochal, which must on ordinary logic leave a great deal of everyday wrongs unavenged. A devotee may not, simply by means of this theory, get rid of Leibnizian worries about the universal balance of right and wrong. Still, it is this reassurance of his subtle presence, which may not take, especially in our fallen times, the form of an avatāra, but of his constant encouragement to the side of justice, which convinces the warrior that Kṛṣṇa is symbolically and effectively with him, that he is fighting Kṛṣṇa's battle.

Ānandamath and the three other later historical novels, I shall argue, should be seen as constituting a single presentation. They ought to be seen as accomplishing just such a condensation of the symbols of praxis (*anuṣilan*), for creation and participation in a myth that is essential as a focus of a large collective historical action.

Bankim's Kṛṣṇa ought to be seen as a Sorelian myth, an ideal condensation of energies, a focus of national-popular mobilization. This is especially revealed by the unremitting use of metaphors and fictions of mobilization in the late novels, its series of myths—the myth of a people, the myth of an army, the myth of a war. And this shows both the daring and the infirmities of his time and his class in its deep ambivalence—the myth is one of political praxis; yet the praxis is only mythical.

4

Imaginary History

History, on an influential recent view, is a series of narratives. Apparently, this is a relatively simple and unsurprising statement. In fact, however, this is one of the strongest and most far-reaching of attacks on the positivistic conception of the historian's art. If history is a subset of all narratives, it displaces the criterion of historicity in a crucial way. Fiction is also narrative. Thus history instead of being distinguished by the trueness of the story, is now distinguished by the storyness of its truth. What distinguishes history then is a form, a way of colligating events; but the events so colligated can be either true or imaginary. This historicity of history does not consist in the fact that it states things which actually take place, but that it sets them out in a particular structure.

In the complex, diverse, noisy discourse of nineteenth-century Bengal, there is above all a new sound. This is the sound of history. The claim of the newness of history is literally true. In Bengali discourse of the nineteenth century 'history' breaks out everywhere, a most remarkable occurrence for a culture which had traditionally treated history as entirely dispensable. Life could be lived eminently well without the trouble involved in giving it a history in the modern sense, and setting up the controls which would ensure that this history was empirically reliable.[1] But it would be wrong to treat this term absent-mindedly and assume that it meant simply a present craving to know about the community's past. What this innocent term means to the first generation of its Bengali worshippers is a complex affair.

History comes to be the great symbol of the new age. It is a symbol with a double significance. In that culture, it is the name of despair and of hope. Yet, this is not a literary paradox; it is true in the most literal of senses. Every colonial intellectual understands the ironical, dual gifts history offers him. The word history, as they well knew, meant two entirely different things: it meant the course of happenings in time, the seamless web of experiences of a people; but its great promise lay in its second meaning, the stories in which what had happened are recovered and explained.[2] Even this dual meaning of history was quite inadequate for all the different things that were done under its name. But this duality of meaning, without any mystification, is constantly exploited by Bengali middle-class intellectuals in various innovative ways. This is their great theatre of truth and imagination, of thought, of social explanation, of philosophy, theology, dreams, political programmes, of their ingenuity, defiance and betrayals. Slowly, history becomes the great terrain of politics. Because history is a way of talking about the collective self, and bringing it into existence.

To say with European orientologists that Indian civilization lacked in linear conception of time, and therefore had no conception of history is misleading.[3] Actually, this does not, in a strict sense, indicate a lack of historical sense. This would imply not a lack of historical sense, but an absence of its precondition. European orientalism made a characteristic exaggeration: in fact, traditional thought had a linear conception of time but with fuzzy ends. If traditional thought had some sort of commonsensically linear vision of time but still lacked a conception of history, this makes the absence more interesting; because the lack is then of the idea of history itself. A culture which had treated history with such indifference, suddenly erupts with a great historical discourse.[4] Western education surely had something to do with this, and the initial search for history must have originated partially from the rationalist imitativeness of the Bengali babu. But it seems inappropriate to attribute this discourse only to cultural imitativeness: since Europeans had a great tradition of history, so must the people educated at Hindu College. The relation between European instruction and what they actually start discussing seems to be considerably more complex.

If judged in rationalistic terms, their efforts often fell far short of the European ideal of constructing a *reliable* account of the people's past; in some other ways, in giving an *imaginative* unity to it, it went far beyond. They invested history, in other words, with a very different function within their cultural discourse, from what it had in European rationalism. Fundamentally, the colonial asymmetry between the European and the Bengali made a simple imitation of the European manner of doing history impossible. To discover the truth of historical objects and connections is the ironical privilege of the subaltern. One of these truths was the constructedness of history, what we today with a language full of finer insults would express as history being ideological; in all history, there is a contamination of interest. Colonial intellectuals of the nineteenth century spontaneously discovered this truth about historical 'truth'. History is not a realm of truths in that bland, straightforward, positive sense; it is a realm of truth established by belief. In other words, history was not a purely academic, but a pragmatic science; its task was not to leave the world as it found it. The past was an image created in the interest of the present. History was, in major respects, the myth of a people, its construction of its self. It was necessary in this urgent sense. What actually happened, what was, so to speak, 'empirically true' was another matter, much less significant. By telling the history of India, Europeans were constructing an essentialist image of a subject people, whose whole history destined them for British conquest;[5] just as by telling their own story of Europe they created a myth of what they were, destined to rule the world through the somewhat long period of preparation since ancient Greece. The lion is always shown as being defeated, Bankim would argue, because it was man who painted the picture.[6] By an extension of this logic, Bengalis initially, but later Indians, must win the right to their own history. They must assert the right to narratives of the self. The empirical stretch of events in history is in any case infinite. To tell a series of infinite incidents is impossible. The conversion of an infinite material into a finite form, which alone could be *told*, is to turn it into a narrative. It appears to the intelligentsia in the late nineteenth century that they must exercise this pre-political right to the narrative of their own people.[7]

Gradually, Bengalis, and later, for identical reasons, Indians in general, became a people obsessed with history, with these narratives and the crucial control over them. They appear to be obsessed with history in every form—history of their own country, history of others who could be as yet only tenuously linked to themselves in a collective we, history of all its periods, of past, and present, sometimes indeed with the history of history itself.[8] Eventually, they would go beyond even this impressive register of real histories and cross over into what can only be termed after Bhudev Mukhopadhyay 'imaginary history'.[9] History means so much because it shows a world in the making, in its contingency, in its open, probabilistic form. It shows not only how the social world became the way it is, but also how close at times it was to being quite different. And it is in this margin of historical possibility that fictional consciousness, like Bankimchandra's, could have free play.

History shows how this world, which so structures human experience and action, came to be. By the same token, it can show how close something that did not happen was to actually happening. By an extension of the same logic, it could show how it was neither unrealistic nor illogical to believe that what was was not something unchangeable, eternally given like nature. History, thus, gave an explanation of the experience of subjection, and also a rational ground for hope. It was possible to change the past in the future, simply by making it the past of a different present.[10]

The Peculiarities of the Discourse of History

The magic that history held for the nineteenth-century intellectual is quite unmysterious. It brought together, rationally, defensibly, two aspects of his life, two poles of his people's complex *mentalité*. They began to write an enormous amount of history, much of which was respectable in modern academic terms, but much besides that was 'fraudulent'.[11] 'Fraudulence', the term Gellner uses so forcefully, captures only part of what is happening, because it implies too strong a norm of objectively true historical accounts.[12] These accounts represent what can be linked to 'the imaginary constitution of society'.[13] Apart from the common classification of different kinds of history according to its subjects, there is another distinction which is, from my point of view, quite crucial. This generation wrote histories of what happened, and also of what never happened; and the interesting thing is that they saw the latter as being part of the enterprise of putting together a historical narrative. They saw through the Macaulayan enterprise of British education in which one was never entirely free of the other.[14] All histories, despite the ideology of positivist objectivity, partly 'describe' some things that did not 'happen'. Indian inferiority, seen racially or otherwise essentially, was not part of what happened; racial superiority of the British was not something that could occur. Yet, through the inevitable transmutation of the historically contingent into the narratively necessary, this was part of every standard history of India. Racial superiority or inferiority cannot be events, but concepts. They do not figure in what is strictly happening at the level of occurrences; they are concepts through which the otherwise meaningless plenitude of events is given some significant order. In this simple and tragic way colonial intellectuals understood a vital truth about the empirical disciplines of society. Narrativization occurs through theoretical structures. There were simply no non-theoretical narratives.

Bankim is especially critical of western writings on India: 'of all intellectual sins (*mahāpāp*) this was the worst, the most irredeemable. Of the many ways of reducing oneself to a verbose stupidity, no other was as effective as the devoted study of western criticism of Indian texts'.[15] He clearly perceived the outlines of what we would today call a definite discursive structure. Surely, European interpreters disagreed among themselves. There were liberal and conservative, English and German, materialist and idealist histories, histories that deplored the Indian's unmanliness and histories that admired him for his devoted spirituality. Such differences, he often says, are not unreal. But in the manner of a tradition of *ideologiekritik*, he finds among such diversities deep structures of common presuppositions. He finds in them that interesting combination of untruth, plausibility and internal consistency which would be regarded by a more radical tradition as the major marks of an ideological structure.

This great apparatus of indological scholarship formed a powerful ideological system, a way of seeing the truth of Indian civilization and history in a manner that conferred legitimacy to something that was apparently unrelated to such scholastic issues: the cultural structure being created by colonialism. These were thus, to use an argument from Edward Said, not really propositions or at least truths of Indian history. Their context of coherence lay elsewhere: they were necessary items of belief in an enormous structure of historical appropriation of other societies that would serve a certain construction of the historical self of Europe.[16] These did not tell any truths about Indian history; yet they contained historical truth in a more complex sense. As displaced truths, truths found at a wrong address, they made sense. Clearly, there was something deeply true about these remarks and representations; but they were not true, as was commonly believed in those times of confident positivism, about the purported objects in regard to which they were made. They yielded, if read that way, some deep truths about the manner and methods of the European construction of its other; what European thought must believe about the other if it was to believe what it wished to believe about itself.

Thus Bengalis, and later Indians, become great history writers, though not necessarily great historians in the more academically restrictive sense. Their larger, abstract concern about history gets articulated in two kinds of history—that of the real and the imaginary; recreation of the past through laborious academic research,[17] and through the different inspiration of the fictive imagination turning towards historical subjects. What they call history consists of these two asymptotic lines. There are many interesting ways in which this peculiar duality is reflected. One of the most curious is that individual authors undertake to write history of both types. Bankimchandra[18] would belong to this rather odd list along with Romesh Chandra Dutt,[19] Rakhaldas Bandyopadhyay[20] and many others. Why should they do both? Should not, according to the great uninfringible classification of rationalist theory, the devotee of facts, of the narrow truth, look upon fiction, imaginary happenings, with an appropriately Platonic scorn? Would not the creative writer be uninterested in the irrefutable presence of historical archives, of undeniable facts unbendable by the imagination? Yet, although they are admirers of rationalism and its ideal of an unsurpassably classified world, with remarkable regularity authors go on doing both, and evidently see them as being in some sense deeply compatible. In fact, sometimes in their thinking the second sort of history appears to be a continuation of the first by other means. Novels are a continuation of history, uttering what history could not. Indeed, we can hardly say that the first type of history is serious and the second is not. Novels had a deep *internal* relation to historical work, for they helped spell the same ideas, only in a different manner or style: they tried to spell, as much as serious history sought to do, a self-respecting relation with the community's past. The Europeans' history of India arranged the events and trajectories in universal history in such a way that Indian subjection to Europe-Britain appeared as a necessity of the historical world. Historical novels and historical treatises by Indian intellectuals sought ways of saying that this was false, by showing the constructedness of this narration, denying its positivist semblance of objectivity, and asserting the contingency at the heart of the historical process. Whether a battle took place on a particular day was a matter of empirical evidence, but whether this showed European superiority was a matter of construction and narrativization.

Fiction which touched upon such questions naturally tended to celebrate history, against the positivistic grain, as a realm of contingency. What happened in history is not always necessary, or if necessary, not in that finalist sense. That was why the writer of history was also, without any internal rupture, the writer of historical fiction. In colonial culture, discourse went through massive but simple ruptures and was subject to great ineradicable repressions. But precisely because these were repressive *structures*, i.e., there was a certainty, regularity and repetitiveness about them, they could be side-stepped by codes. Conclusions, theoretical statements, speculations about history's non-narrative meaning, got displaced and erased out of the academic part of historical reflection. These sought refuge elsewhere, and were displaced into this other discourse of fiction, which followed it like its shadow. And if the audience was to make out its meaning properly, they had to put its two disjointed parts together, only then would it begin to speak. History's hidden discourse, its great wealth of conclusions and corollaries surfaced in this other territory, across the frontier of scientific realism. Both the writer and reader were therefore creator and taker of not a text, but of a discursive order. It is in this other, necessarily parallel discourse that one could hear the hidden and exiled truth of the straightforward texts of history.

Such homology, proximity, mixture, is too numerous to be exhaustively recounted. This combination was given, in the case of different individuals, different temperamental emphasis; some were more story writers, others more historians. Yet what is remarkable is that most intellectuals attempted both. Bankimchandra, in irritated and excited intervals between composing the imaginary history of an as yet unnamed people (we shall see why they are unnamed), came back obsessively to an attempted history of the Bengalis, or of the Hindus, or of the Indians, on the ground that in order to graduate from being merely Bengalis into human beings, his people needed this obligatory narrative.[21] That he could not complete these tracts should be traced, I suppose, to a reason deeper than sheer lack of time or interest, to a lack of faith. It is to be traced to his dark certainty that an academic, rationalistic history of Bengal would not provide him with what he wanted—a series of symbolic events of defiance, of great acts which this people could be exhorted to remember, and when the time came, to re-enact. Bengali history, from this point of view, was disappointing;[22] but this was less true of a possible history of the Hindus, and even less true of the Indians,[23] which showed different payoffs for where one decided to draw the line of one's belongingness. There is, remarkably, a great tentativeness, about this people (*jāti*) Bankim would like to call his own: he seems unable to make up his mind about which people to belong to, and several choices seemed open to him. Often, this 'natural' community, or more appropriately, this 'natural' *we* opens up its boundaries and moves towards a greater and imaginary *one*.[24]

Interestingly, those who write such imaginary history often take advantage of a conceptual indeterminacy. Human beings do of course live in communities, and the language of every group has words or concepts which designate the community to which an individual is supposed to belong. But often such language works on a peculiar presumption of self-evidence designating that 'community' by a generic term. This appears to be true of languages in societies which have no consciousness of either the optional or constructed character of communities, or the layered nature of the community in which people live, in which, depending on the context, one can designate the neighbourhood, the region, the nation (if one is available), the religion. In such languages, the conceptual edges of the term 'community' are fuzzily drawn. Indeed, one could speak of a double fuzziness: (i) the term is indeterminate between its varying, competing meanings; besides (ii) even when given a single meaning, on a single level, its boundaries remain to some extent unclear.[25] They are not used to the modern manner of living with clear enumerated definitions of community, in a world which is, from the modern point of view, very unsatisfactorily classified. In times of historical conflict these crucial undetermined frontiers get sketched out, and put in place. This indeterminacy figures in Bankim's work in two ways. It is clearly reflected in his uncharacteristically ambiguous use of the word jāti for all these communities, starting from castes, to a regional people, to religious communities, to the nation. But it is also used in a second way: fiction writers used the fuzziness of this idea of a community to give their audience a community which had not existed before, by gradually conceiving a community called the nation, or selecting the appellation of the nation for one of these communities.

The complementarity of the two dissimilar narratives of history is illustrated by the literary careers of many Bengali intellectuals, but the two obvious examples are Bankim and Romesh Chandra Dutt, the author of the *Economic History of India*. Dutt produces one of the seminal texts of anti-colonial political economy, yet he writes, alongside, his two famous historical novels with the clearly counterpointed titles: *Rājput Jīvan Sandhyā* and *Mahārāṣtra Jīvan Prabhāt*, strange generic titles for what are essentially romantic stories. This extends Bankim's idea of breaking up the identification of the collective self as Bengali, and opening it outwards, letting the history of other peoples figure in the narrative of the Bengalis' *we*. These are therefore crucial moves in the direction of an authentic nationalist history of *India*. True, British historians had earlier written histories of India. But, for them, India was essentially a territory, the colony British imperialists had occupied, interesting for the history of its past. What these colonial authors attempted was a very different history—in its manner, aims and even its object. They enter a history of India not externally, but internally, through the process of making an India that was not there. If we analyse Bhudev Mukhopadhyay's phrase, *swapnalabdha bhāratvarṣer itihās*, the qualifier *swapnalabdha* relates to both terms, not only to *itihās*, but also *bhāratvarṣa*. It is not only the history (narrative) which is a product of a dream. India, whose history this purports to be, is as much a creation of dreams. From a geographic entity, India is becoming through such tragic moves of inclusion, an imagined community. It is wrong to say that this history is a different history of the same object; it is, rather, a history of a different new object, an object which is interesting for the history of its future. It becomes in Bankim, probably for the first time in Bengali literature, something emphatically other than simple neutral territory, a profane space, but sacred 'ground' (with all the great complexity of this metaphor) of a community. Space is invested with sacrality in a literal sense making transfer of a moral language possible. It was not something which was fit to be geologically surveyed, but to be offered a political form of worship. From a neutral space, India becomes an evocative symbol, female, maternal, infinitely bounteous, invested with the complex and convex symbolism of the feminine in the Hindu tradition—a sign simultaneously of vulnerability and invincibility. Remarkably, it was a little inappropriate to speak of her as a symbol, because she was also, for them, literally, unmetaphorically, this land, and therefore not representing something other than what she was.

Dutt and Bankim thus illustrate in their perfect complementarity the complex dualism of the new historical discourse. Each history is incomplete without the other. Bankim cannot turn Bengalis into full human beings without conferring history on them. As much, Dutt cannot convince his audience of their collective ability to defend, to defy, to act for justice, only on the rather prosaic basis of his economic history; he requires an imagined counterpart. However, it is not imaginary in the sense of being illusory; on the contrary, it can convince simply because it is, in a different mode of writing, something fundamentally true. The novels are so important precisely because they state elsewhere what can only be called the unasserted rhetoric of the *Economic History*. Surely, Dutt in his *Economic History* speaks, twenty years after Bankim, in truly moderate tones, of measures which regretfully 'weakens the Empire and impoverishes the nation'; yet, within this guarded, upper-class moderation, something of enormous significance has already happened. Something called the 'nation' with more of less clear boundaries has come quietly but ineradicably into existence.

Besides, there is another way in which the two sets of writings have a common point. The political economy of the *History* polemicizes against the apologetic political economy of colonialism. In colonial culture, there are also large cultural constructs of the 'effeminate Hindoo' or the Indian permanently fearful of battle, an eternal slave to the arms of invading peoples, drowned in sensuality, incapable of energetic action. Imaginary history, or historical novels contest this stereotype. In battles which were probably disastrous routs, there could be instances of individual heroism. In a long history of subjection there are great episodes of real resistance. Fiction simply invents imaginary subjects to confer such real predicates on them. Fiction becomes in a strange reversal the repository of truths which might otherwise be lost. Historians of a more academic bent constantly do something comparable. Radical writers of one generation recounted the resistance of communists within the general betrayals of bourgeois nationalism; subaltern historians recapture the defiance of the nameless and unchronicled poor within the history of elite collaboration. Such stories obviously are not falsehoods simply because they 'go against the grain'; these are precisely history within history. The truth that nineteenth-century fiction sought was largely similar.

If we have to read individual texts, we must therefore subject them to the kind of reading that Freud recommended for dreams. Displacement, condensation, masking and metaphors structure these texts like that, for dreams are, like the thoughts of the unhappy consciousness of colonialism, decentred, a combination of desiring, saying and hiding not only from the world but also from the self. These are essential for understanding how the Calcutta babus, with their friends and relations, come to be a 'people' called Bengalis, and how this chauvinistic, pampered and self-indulgent group concluded that they must be Indians.

A Critique of Essentialism

Historical and fictional narratives are allies in the same enterprise, because they share a common adversary, and this adversary speaks primarily in an essentialist language.[26] Some forms of essentialist ideas were of course very common, and were the general currency of British Indian education. Partha Chatterjee has recently shown how much of colonial administrative and political thought displayed an *essentialist* form.[27]

Essentialism, as he points out, could assume two distinct forms. A first form would deny the existence of history to a people, simply because the rhythms of its evolution might be quite different from what is common to Europe. A more sensitive form of the essentialist argument would not deny history in that deep sense, because of evident restructuration of a society's arrangement over time; it would however find ways of nullifying this surface level change by pointing to a supposedly stable structure of characteristics which are transcendent, to which history can only give form, can variously refigurate, but cannot really alter. Clearly, this metaphysical register, or essence is simply an abstract construct taking the most general features out of history selectively. But such forms of thinking are only too common in colonial India. Examples of the first form of essentialism were the stereotypes of Indian effeminateness (and it is interesting how, despite some ambivalence, Bankim turns the feminine into his constant symbol of power),[28] of their depending on cunning rather than force in dealing with the enemy, the unchangingness of Indian society, of either the exclusive licentiousness or the exclusive prudery of the Indian civilization. Assuredly, essentialism had other, more retail applications. While it was impossible to find obedient equivalents to the economic, political, social, it was, conversely, equally difficult to find translations of dharma,[29] evidently not equivalent to religion, artha, immeasurably wider than the mere economic,[30] and kāma, very inconveniently related to love, passion and desire.

A more sophisticated form of essentialism treated historicity somewhat differently. Bankim is acutely conscious of this version, as is apparent from his constant abrasive references to Max Müller, the most eminent perhaps of the spokesmen of the second form. This form of essentialism did not deny the phenomenal form of history, that societies of the Orient too underwent significant change in historical time. But this change simply unfolded, restructured and reformed a fundamental essentialist alphabet. Religious doctrine in India would not be seen as unchanging. Rather, the great historical interest of the Indian civilization would lie precisely in the enormous variety of religious life-forms. Nor would these forms of religious doctrine be seen as historically static; historians would often analyse and chart their historical trajectories with great care. But these allowances for historical change would still be formed and held within a general premise that Indian civilization was essentially spiritualist, and consequently, the evidence regarding the history of this civilization would be run through an elementary censorship or erasure.

Some varieties of essentialism were very common, and formed the general diet of western style higher education. Western education was a deeply contradictory experience in many ways. It involved exposure to the rationalist stock of knowledge, an entry so valuable to his generation into a disenchanted world, a world of causalities, no longer of meanings. And this world of causalities, its inner algebra, its methodology, was open to one by the simple fact of being human: rationalist knowledge, seen through this generous humanist construction, was a great principle of enfranchisement. But the identical processes of western education exposed these intellectuals to, if not forced them to accept, the idea of European superiority on some subterfuge of race, or culture. The commonest form of this essentialism about Europe was of course the triumphantly linear, cheerful history of Europe, which constantly gerrymandered the boundaries of this privileged continent to keep up the optimistic story of its culture, to confer on it a retrospective inevitability of movement from Greek antiquity to exalted imperial status in the nineteenth century. Bankim has an acuter sense than most of his contemporaries that this history is equally constructed, that western rationalism, in its colonialist form, cannot afford to be truthful about its own history as much as about those of others. It is not only oriental history which is an imaginary history in a derogatory form; its account of the European record is equally an imaginary history, though there it tells a positive fable. Occasionally Bankim is surprisingly close to asserting that history is not a field of 'knowing that', but of 'making sense'; that history has no purely objective substratum, but is essentially this construction, this putting into useful form, of the malleable infinity of the past, of the great, complex, always unexhausted, inexhaustible resources of events which are available for recall. But his eventually rationalistic intellectual formation pulls him back from the brink, from the strain and exhilaration of living in such a factless, unobjective, underdetermined history. His deep sense of subalternity, of an unjust principle of power overriding the enfranchising principle of reason, however, constantly tempts him. Insight into the fact that the European history of Europe and its other are narratives, insecurely different from fictions, enables him, and authorizes his generation to do a negative construction on the same principle. History is always deeply, ineradicably fictive: and therefore explicitly fictive principles of organizing a history of one's community could not be a priori condemned as illegitimate. Bankim's language rings constantly with such declarations of symmetry between fact and fiction. He has already announced this equivalence, and created this space in *Kamalākānta*. This would be, strictly speaking, on the same level as rationalistic or Macaulayan history, because 'that is fiction too' (*tāhāo novel bate*).[31]

Their mistrust of history—its claim of objectivity and rationality—shows outlines of an interesting line of thought about what history is, how it is organized, why people so insistently give themselves histories of themselves *and of others*. Representations create a sense of power. History is an empowering discourse, it gives reasons, justifications for important practical initiatives: to fulfil its appointed purpose, it must become a history of its people *in the world*. Thus a real history of its own people, of its collective self, must also include historical narratives about others—those who are different, strange, unfamiliar, subordinate, threatening. The discourse of history is composed then of two distinct but related elements. The first could be called the level of empirical *facts*, the solid bits of information—checking and getting them right. But this level of history, its first-level empirical stratum is not history in the proper sense; because history, properly so called, is an *order* of facts, and this order cannot be produced out of the facts themselves. To stretch their argument analytically, even the chronological order, the minimal, the most indubitable, the most elementary of orders, depends on non-factual judgements. Chronological telling of history is committed to recounting the next fact in the chronological order, a *y* which follows a chronologically prior *x*. Yet the judgement that *y* is the event that falls in the same class as *x* is one that is not made from within grounds that lie entirely inside *x* as a fact. It is a narrative judgement which can tell the historian that *y* is an event of the same class as *x*, and is therefore the eligible candidate for the empty place after *x* in the narrative that is being unfolded. Colonial intellectuals argue that history could be false or ideological at both levels: the elementary determination of events could be faulty and often was; the narratives could be ideological, for all narratives are. Therefore, the point in history is to choose between the narratives on offer not because of their truth, but because of their ideology. More reliable information on a battle between the British and the Marathas might or might not dispel confusion or interested misreporting. The battle however would always be available to a second order of discourse, however reliably the numbers and dispositions are determined. It can always be seen as one of the last signs of resistance to the triumphant empire, or one of the first battles of resistance against it. Events of 1857–59 have in fact undergone similar narrative transformations: from the last desperate kick of the effete Mughal empire it has slowly become the first war of independence.

The history of the British went, opportunistically it must have seemed to this embittered generation, from nineteenth-century Britain back through the history of European renaissance—not all of which was British—even more interestingly straight to the antiquity of ancient Greece. Gerrymandering of historical identity of this kind was too blatant to go unchallenged or unridiculed. Understandably, the strategy of critical intellectuals was to point out the untenable character of this cultural ancestry. Many of them pointed out that the cultured Englishman's presentation of his forbears among renaissance Italians and ancient Greeks involved neglect of the more racially intimate Celts.[32] By the middle of the nineteenth century the battle of the forbears was well and truly under way, each side pointing to the weaker elements in the others' construction of their beautiful historical self. Thus, while Max Müller warned that the sacred texts of Asiatic religions must be seen with their share of horrific and silly beliefs,[33] Bengali writers took delight in a comparative cultural chronology of the rulers and the ruled. Not surprisingly, references are extremely common to a time in which, though ancient Indians were perhaps writing texts in which much that was valuable was mixed with much that was downright silly, ancient Celts were engaged in delights of a predominantly non-philosophic kind. Max Müller's insistence and scholarly anxiety that ancient man should not be glamourized, was turned around, and implicitly told that the points of civilizational comparisons should be temporally synchronous, satisfaction drawn out of an implicit comparison between ancient India and modern Europe was deeply anachronistic and methodologically flawed. It is important to remember the great asymmetry in this debate: European ideas were more speaking than spoken to: the debate for them was not truly dialogical. For Bengali intellectuals too it was monologic in a different sense: they listened to the discourse of Europe about Asia, analysed and studied those constructions carefully, but inequality of power stopped the discourse from being dialogic, because there was no way in which European scholars could be made to take notice of what their garrulous subjects thought of what they said. This strange parallax contributed to the curious structures of this debate. It constantly tried to prevent changes in historical geography, keeping the frontiers of peoples straight or constant through historical time, preventing England from including Greece and Rome within itself. Alongside, there was also the gradual radicalizing of this debate and its arguments: not merely were two contemporary peoples seen to be involved in this contest, but two pasts, their cultures, literatures, even their literary characters.[34] Few other arguments could be as total as this, and as essentialist.

Small infelicities of this kind—an occasional geographic opportunism, or accounting for the few centuries of the middle ages—could cause no embarrassment to the cheerfully linear history of European reason which saw the whole of European history as a preparation for the European mastery over the colonial world. No structural irrationality of other societies could withstand the principles of the enquiring reason. Ideas of this early anthropology asserted that all historical societies could be arranged in a rationally defensible hierarchical order; though, the classification of 'irrational' societies tended to be somewhat perfunctory when compared to the great red line separating modern Europe from the rest of time and space.[35] Interestingly, cognitive relations between these societies were one-way—while higher order societies could enquire about and rationally explain or understand lower ones, its reverse was not true. Lower societies were opaque to the concepts of their own inhabitants, and gave up their structural mysteries only to knowledge produced in a higher form, anthropology in particular. The European side of the colonial discourse is therefore full of amateur and professional ethnologists who had undertaken, for the abstract love of knowledge, to demystify the mythology and religious beliefs of colonial peoples. It is not surprising that the intended beneficiaries of this operation often resented their labour.

Some versions of essentialism implicitly used the dualistic tradition of European philosophy to patronize Indian culture. The material and the spiritual were two transcendent categories in which all human things could be classified, and although the Europeans were better at material enterprise and the science of nature, they could always learn from the spirituality of ancient India—once again a curious transaction across historical time. Bengali intellectuals often resented this form of cultural appreciation; and no doubt, this would have made European orientalists wonder about the perversity of this intellectual group. For apparently, they could not take European admiration of their culture any better than they could European criticism. To the subordinate group and its intellectuals, the surprisingly total experience of colonialism invested everything with a political meaning, deeply felt precisely because it went unannounced. This appreciation of the spiritual excellence of Hindu culture and philosophy was read—correctly and wrongly—by the local intelligentsia for its 'absent' political implication. As material enterprise ruled the world and European supremacy in technology undergirded the whole structure of colonialism, this provided, just as much as did the first theory, a justification of colonial domination of Asian societies.

It was possible to break out of such mental prisons only by thinking *historically*. Different writers inflected this injunction in different, individually specific ways. But there was an atmosphere of hunger for history (*itihāsbubhukṣā*)[36] as Tagore put it. One line of Bengali intellectualism implicitly claimed that Indians were always superior to Europeans in spiritual matters; although they required much European succour to claim this superiority. A conservative Brahmin reasoner like Bhudev Mukhopadhyay could argue[37] that the principle of Indian history was fundamentally different from the European. 'It is essential to reject the superstition that the history of all countries will be similar. An expert in the biography of Rothschild might, in the case of Christ's life, call for his financial accounts and office records; and if he does not get them, he might evince contempt and say what is the use of a biography of a man who did not earn a penny? Similarly, those who feel disappointed when they do not find royal ancestries and chronicles of victories and defeats from India's official archive, and assert that there is no history where there is no politics, simply look for the wrong kind of thing'.[38] European history expresses the principle of conflict of self-interest; Indian history expresses the principle of finding unity among dissimilars.[39]

Although the response to European essentialisms led to such popular and poetic essentialisms of a negative sort, some intellectuals saw history as the answer. To get out of such traps, it was essential to see Indian subjection and English supremacy as historical. This was why history was so dear to the oppressed. History thus revealed to colonial intellectuals its great philosophical secret: history alone allowed us, provided us, with a category to break out of the forms of essentialism.

Even on that day, even under that dustladen sky the flow of birth, death

> happiness and sorrow which flows through the homes of the village—for men, even though it is covered by dust, is the principal thing. To a foreign traveller however this storm is the primary reality: its cloud of dust covers everything else from his eyes. For he is not inside the home, he is outside. Thus, in the histories written by foreigners, we get only the tales of this dust, of this storm; nothing about the home. Reading that history it appears as if this India did not exist, the only thing that existed was this thundering storm of the Mughals and Pathans spinning its column of dry leaves from north to south, from east to west . . . But when foreigners existed, the natives existed as well . . . Our deep links are with that India which lies outside the textbooks of today'.[40]

It was history of a new kind which was to end this servile and misleading recounting of misfortunes.

Irrespective of which mode an author prefers, the materials for this historical discourse are diverse. They are drawn from three kinds of sources at the least. The first stratum is constituted by the events themselves, which for this first positivist generation, is a pleasantly unproblematic scientific concept; though, surprisingly, on occasion, an individual like Tagore might state as a poetic question some doubts which appear interestingly sceptical of such certainties. History is seen overwhelmingly as a social and collective enterprise, something that the historian is doing as a representative of his community. It is not a matter that one enters by individual curiosity. This is indelibly marked on the rhetoric of this historical discourse: it is nearly always that the subject of this history is written, even grammatically, as 'we'. As a consequence, as they are the victims, not the victors of their modern history, they are far more intuitively conscious of the problematic character of the pre-narrativized event, a pre-theoretical fact, a pre-constructed incident, which has happened, but is waiting virginly for its true account. Their sense of subalternity immediately problematizes the concept of a pre-theoretical fact. They come to feel with the sharp sensitivity of the insulted that there is no fact in this ideological terrain of history which is not part of a narrative design, which is not already contaminated by a larger governing story and its hidden prejudices.

A second type of material for the construction of history comes from literature, or to use a more fashionable term, the high discourse. Recounting of actual events is rejected on two grounds. First, the little that is available of this history, is the history written by the unsympathetic, often aggressive alien.[41] Histories of Bengal written by Englishmen[42] are actuated not only by a curiosity about the Bengali past, but an impulse to justify its present. As Tagore declares, this is a sort of history of India in which Indians are absent.[43] Indeed, this employs a line of reasoning fairly common to all Indian writing: one central difference between Indian and European history was seen to be the relation between society and the state. The state, the political order, it is commonly believed, is organically related to the social form; so that the state has the ability to deliberately direct and restructure social orders.[44] Consequently, political upheavals usually have social consequences; and conversely, if the social order is to be altered, it requires an assault on the state. Indian history, Tagore and others argue, is marked by a marginalization of the state, of the political order at least in its narrow sense. Tagore turns this thesis to make his fundamental critical point—that this history of India is inconsequential, if not false. Political history is not the means by which the secrets of Indian history can be revealed.[45]

This could be a prolegomena to a seriously painstaking social history. But that line is not taken up. Instead, in a famous and illustrative essay, 'The course of Indian history', Tagore starts reading the 'documents' of this history in its literary and theoretical texts.[46] The history of India is inscribed in archives of a different kind, the archives of its intellectual history, the *Rāmāyana*, the *Mahābhārata* and the Buddhist scriptures, for these reveal the basic trends in Indian history: they are less factual records, as dominant and popularly accepted interpretations of what happened.[47] They operate not at the level of 'mere fact', but of historical 'truth'. In this enterprise of history, these epics make constant appearances, and they are accompanied by constant and entirely absurd comparisons between masterpieces of Indian and Western literature. But this absurdity is precisely the problem: why should perfectly sensible and highly cultivated people wish to decide with finality whether Kālidāsa is superior to Shakespeare? It shows again an act of displacement. Under the excuse of comparing Shakespeare with Kālidāsa, they were making statements about something quite different. These are attempts to pay European rationalists back in kind, to make an egregious statement of the superiority of Śakuntala over Miranda and Desdemona. These figures are called up as reserves to fight a battle whose total, all-pervasive character is indexed precisely by the absurdity of the weapons by which it is fought.

The third type of materials are of course the rationalistic histories of India. These are attacked by all manner of arguments—from simple falsification, to misinterpretation, to more sophisticated problems of hermeneutic entree into a culture. But Bankim quite often also uses an unanswerable means, his banter, making hilarious travesties of European scholarship of Indian literature. To this naturally the European writers, or the conveyers of their views in India, could make no answer: for travesty narrowed down the orbit of discourse; it internalized the discussion among Bengali speakers, and effectively excluded the European.

From their point of view, history is an indispensable implement of living, for people's coping with their social world. But this history is not the one that historians make. Indeed, there is a connection between the historians' chronicles and what people believe to be true of their common past, but it is more complex than is usually believed. Popular beliefs do not obediently follow the historians' record. It is not so resourceless and passive. Often it would take the basic material from the historians' record, but construct out of it something entirely its own. In such transformation, in this historically crucial bending of the actual in the direction of the plausible, the popular mind often takes help from the writers of fiction. It is important to understand the precise nature of this transition from the actual to the plausible, and the transition from intellectual to popular beliefs. The historians' grounds are what is archivally credible. The principle of popular imagination is an enhancement of the present.

Evidence seems to point towards something like this happening in nineteenth-century Bengal: this could be called a reception theory of historical narrative.[48] There is no end to the odd and downright silly things that people can believe about their historical past. Middle-class Bengali intellectuals found ways of believing, against formidable evidence, their indisputable descent from the Aryan writers of the *Vedas*. Although race was a stable biological feature, his complexion had only somewhat darkened, his structure somewhat shortened, his nose lost some of its aquilinity due to the ravages of tropical weather. Otherwise his racial purity was self-evident. Sometimes their beliefs about what happened in the past were wrong. But even if their image of what the Aryans did was correct, what was doubtful was the claim that it was *their* past. In the business of stitching together Bengalis with a certain past, they were also doing something deeply understated—stitching Bengalis to others in the present.[49]

Bankim is particularly important in this search for a collective self, for he made a powerful proposal to his contemporaries about what sort of a collective memory they should have. This related to a shift in Bankim's own development. After *Kamalākānta* his reflections seem to have undergone a change. *Kamalākānta* had posed a recognizably historical question.[50] Later, he speaks of history more often, less ironically. Laughter is still sparkling in what he writes: there is still some sense that the world is an object of irreverent mirth, but the distribution has changed. He is not writing in order to evoke laughter; he is laughing in the intervals of doing something else.

Along with humour, history is the great theatre of alternatives. Bhudev Mukhopadhyay invented a famous phrase, used as a title of one of his weak novels— *Swapnalabdha Bhāratvarṣer Itihās*—a history of India revealed or found in a dream, a dream history of displacements and condensations, which spoke in a language of fantasy about the most unforgettable problems of their everyday. Bhudev had invented only the bare idea but he lacked the power of imagination to make proper use of his discovery. It was Bankim who started creating in his novels the theatre of utopia and wish fulfilment, in the realm of his artistic sovereignty where he could legislate with impunity against the known historical record, an alternative history of India. It is in his later novels that he begins to choose his nation.

The Discourse of Factual History

Like most of his contemporaries, Bankim's search for history began in a curiosity about the empirical past, driven no doubt by the belief that to live without knowledge about the past was uncivilized. But this programme was one of the pursuit of an elite *bildung*, of becoming properly cultivated which included being properly informed.

It is not surprising in view of this that the early, intense consciousness of history, 'this hunger for a history' was plagued by a curious indecision. Of course, it was clear to these intellectuals that they must have a history. But who were this 'they' who must have a history? Where, and how, do they draw the lines around themselves? The factual history that Bankim initially sought was out of an immediate consciousness of the group, or people who faced the British in the colonial situation, and needed an intellectual solution for their indignity. They were 'the Bengalis'. Certainly, there was at work, even in this first elementary move, a certain mythicization, and as always happens in myths, an appropriation. The indignity of colonialism was felt very differently by other Bengalis besides the babus. There was no obvious reason why they might agree to constitute a people, indivisible and one, along with the disgruntled bhadralok of Calcutta. Reasons for such stitching together were provided by the essays on equality[51] and, more particularly, on the 'peasantry of Bengal'.[52] In an exchange of agonies, indignation and humiliation, the loquacious babu was made to give their sufferings a voice; the blurred mass of the non-babu poor would own the very different sufferings of the babu. This, if accomplished, would yield an imposing combination of a people of Bengalis, to be seen now as indivisible, whole and organic. '*Bāṅgālīr itihās cāi, nahile Bāṅglaī kakhana mānuṣ haibe nā*':[53] the Bengalis must have a history, or else, they would never become human beings. It is as simple and fundamental a relation as that. Part of the answer to our riddle is hidden in the sentence itself, in the invisibility of its grammar: in the violent present tense in which it speaks. This need for history is not the academic's or historian's need; no cognitive need can produce such violence of language: '*sāhebrā pakhī mārite jān tāhāro itihās likhita hay, kintu bāṅgalār itihās nāi*'.[54] Europeans (but here used in the narrower sense of Englishmen) create history even when they go duckshooting, but between themselves, a whole nation of Bengalis have not achieved a history. Clearly, there is a typical inverting of sequences here: the 'nation' of Bengalis are considered to precede their history: the community is pre-existing. It exists in a 'historyless' state, and requires a history for itself. The actual relation is as we shall see, the obverse of this: it is the commonness of this narrative which would constitute the community.

Bengalis need history, not because they are assailed by an urge to know what happened in their past, say, how much tax burden was there on the medieval peasantry, or how many years the Pala Kings had ruled, or, in M.N. Roy's words, exactly how many sacks of *kishmish* the great Aurangzeb had consumed in his illustrious life.[55] Evidently, they can exist in passable happiness without knowledge of such technical stuff. Indeed, they require history for quite a different objective—becoming human beings in the present. What a combination of ambiguity, complexity, anxiety of the elite intellectual! A history is required for the Bengalis not in the ontological sense of history, which Bengalis, without much special exertion on their part, surely had. What they require is an *account*, mortifying and uplifting; they do not lack the events in their past, but the narrative to put some significant order into them. What a strong proposition of the connection between narrative and collective identity (humanity)!

The truth of history that the Bengalis need is related to their achievement of self-consciousness. Consider the asymmetry between the Bengalis and the British inside this contemptuous sentence. The British cannot help creating history, they create history in spite of themselves, absent-mindedly. The Bengalis cannot create history despite the historical exertions of an entire people. That the British and the Bengali are held in a relation of subjection and domination appears in some way connected to their relation to history. If we translate the two simple codes in the sentence, the English rule—they have power, because they have history; the Bengalis are ruled, they always lack power, because they do not. History has come to be what myths explicitly are—stories that societies tell about themselves whose points, inevitably, are not in the past but in the present.

In a great many fragments, Bankim returns to the question of writing a factual, scholarly history of 'his people', the Bengalis. All of them are marked by a peculiar fragmentariness, which seems to me to indicate something more basic than simple lack of time. Repeatedly, the resolution of writing a history of his people is taken and forsaken. This is, I suggest, because this history is not able to deliver what he wanted, a history, however disjointed and idiosyncratic, of defiance, occasions of rebelliousness against invasion or injustice, metaphorical counterparts in other words for the predicament of modern Bengalis. One cannot speak of the ancestors of the modern Bengali people till fairly recently in historical time.[56] The advantages of this discovery, the compensations of the sacrifice of ancient glory, are considerable. A great advantage is spelled out by Bankim himself in a wonderful passage warning against the dangers of anachronism:

> When we settle down to read the history of Bengal, we read that the Pala and Sena dynasties were rulers of Bengal: Bakhtiyar Khilji conquered Bengal; the Pathans came to rule Bengal etc. etc. All these are errors for, at the time of

> the Senas, Palas or Bakhtiyar, there was no kingdom known as Bengal. There was no other name for this Bengal of today. The Senas and Palas were rulers of Gaud; Bakhtiyar Khilji conquered Lakṣmanāvatī. Gaud or Lakṣmanāvatī are not ancient names of Bengal. No people (nation) called Bengalis inhabited that land. . . . First was religious unification, linguistic unification, later this came under a single, unified rule and turned into modern Bengal. That political unification has happened comparatively recently, during British rule. The Muslims were never able to unify the land of the Bengalis. The Mughals did to a large extent, but even they could not become the rulers of modern Bengal. Thus, the sense in which Greece has a history, or Rome, in that sense, Bengalis have no history.[57]

The obvious advantage of this line of thought is that if the modern Bengali people did not exist at the time of the early Muslim conquerors, the claim of some British historians, but more than that, the essentialist stereotype of a people forever subordinated, falls to the ground. It was widely believed, on the authority of British historians, who take it from a Muslim historian writing sixty years after the event, that Bengal, at least, the capital of the effete Sena ruler, was conquered by seventeen cavalrymen. To Bankim, this appears to have been an obsessively symbolic myth, to which he relates contradictorily, using it sometimes to berate the babus, destroying its evidence, angrily sometimes, but always with great passion. He asks educated Bengalis, converts to western rationalism, how do they believe such a rationally, materially untrustworthy story, and answers: 'This belief has no other reason, except that the British have written their histories in English adopting Minhazuddin's story. To read history makes it possible to get jobs. Why should you not believe?'[58] His historical scepticism comes to touch even the edge of the founding event and myth of British rule. 'It is said in history in the Battle of Plassey a handful of Englishmen and their native soldiers destroyed thousands of native soldiers and won a wonderful victory. This is pure fiction. There was no real battle at Plassey. There was a mere pretence'.[59]

To state this in other words, these are myths required by conquering elites in order to spread a native superstition about their invincibility. These myths are to be destroyed by rational history. 'There is no history of Bengal: what exists is partly fiction, partly the biographies of the alien, insubstantial, oppressive rulers of alien religions. We need a history of Bengal, or else Bengalis are doomed'.[60] This line of reasoning about history is a result of disappointment with the Bengalis' political history, and is scarcely different from Tagore's more poetically expressed plea in favour of a displacement towards a history of the society.[61] Political history is disappointing, because Indians have a long history of subjection; for Bankim, except for short interludes, this includes the time of Muslim rule over India.[62] Indians, Hindus, Bengalis were subjected so easily because of the lack of physical strength—in the essentialist myths of colonial Bengal.

Only part of this, about the Bengalis, is true. Tales of the military prowess of Bengalis in ancient times are simply self-indulgent rumours.

Although the Bengalis have always lacked military and physical strength, this is not true of Indians in general. From the times of ancient Greeks, most invading generals have commended the military courage of the Indians. Even the English, who always speak of the effeminate 'Hindoos' have suffered military reverses against the Marathas and the Sikhs. Why then did the Indians, or Hindus, fall under subjection? It is difficult to establish the military glory of the Hindus because of their peculiar forgetfulness, because of the lack of chronicles. Chronicles which do exist are written by others, which were essentially, as shown earlier, part chronicles, part narratives of self-mythicization of the rulers.

However, if Hindus and Indians are not the effeminate race of colonial myths, in a sense the question becomes more difficult to answer. Why are they so vulnerable to others, why can they not overcome subjection? Answers to this question must lie, logically, in a social principle.[63]

Indians, he declares, are naturally (*swavābatai*) devoid of a spirit of freedom. Ordinary people in India never felt that they would allow themselves to be ruled by people other than of their own race or community. There was an equal likelihood of good or oppressive rule by own or alien rulers. Therefore, let the ruler try to defend his kingdom, or fail; 'we would not hurt our finger' for either of them—this appeared to be the primary line of thought among Indians. Indifference to independence in this sense seems to be a long inheritance of the Hindus, not a failure of modern generations.[64]

Bankim's explanation of this indifference is a combination of a climatic theory of the type of Buckle, combined with a theory of religion that strongly resembles later arguments in Max Weber.[65] The entire drive of Hindu philosophy and religious doctrine was towards renunciation of the world. Buddhism, though dissenting from classical Hinduism, was no different on this point. This world-renouncing doctrine of the Indian religion made exertion for the material prosperity look unimportant. If a people do not believe in material comfort, this will lead to a lack of enterprise. Indifference to freedom is a part of this lack. When the king or the general was killed in battle Hindu armies fled because they saw no point to it any longer. 'For over three thousand years, Aryans fought against Aryans, Aryans against Delhi, Delhi against Lahore, Hindus against Pathans, Pathans against Mughals, Mughals against the English—everybody quarrelled with everybody else and destroyed the land by the fire of war. But all these were wars between rulers; the general Hindu society never fought against anyone or on behalf of anyone. So it can be stated that Hindu kings or the kings of Hindustan were repeatedly defeated by others, but not that the general Hindu society was conquered, for Hindu society as a whole never fought against any aliens'.[66]

This is small consolation for it simply rhetorically protects Hindu society from historical defeats. 'This has brought us to the second reason for the long history of dependence of the Hindu community (Hindu jāti). That reason is the disunity of Hindu society, the absence of an emergent sense of community/nation, want of a national spirit or whatever else'.[67] The primary difficulty with Indian society is its persistent segmentation, 'In India, there are many peoples. These are due to differences in language, of lineage, or religion. Bengalis, Punjabis, Tailangis, Maharashtrians, Rajputs, Jats, Hindus, Muslims—among them who would unite with whom? If they have religious unity, they lack ethnic unity, if they have ethnic unity, they lack unity of language, if they have unity of language, they do not have unity of residence'.[68] There are cryptic references to three episodes in Indian history—of the Rajputs, especially Mewar, Shivaji and Ranajit Singh, symbolic episodes which are celebrated in Tagore's nationalistic poetry much later. Finally, there is an amazingly double-faced passage; the English have given Indians invaluable education; they have taught the Hindus what nationalism means.[69] That no one should miss its point, there is a footnote which says: 'in this essay the word *jāti* means nationality or nation'.[70]

If Indians really learn what nationalism is from English instruction, ironically, its main political target would be the English empire. But this leads to a mixed, unclear, uncomfortable discussion on 'India's freedom and dependence'.[71] Several interesting problems arise in interpreting Bankim's paradoxical and tortured reflections on the question of independence. Quite obviously, there is a sort of pleasure in even uttering the forbidden word. But the indeterminacy of the collective whose agency this independence would be in some distant but illumined future plagues his thought. It is important and desirable for some unspecified people, the 'we' to which Bankim belongs, to achieve independence and liberty; but this 'we' remains ambiguous. Indeed, the exact character of this ambiguity is interesting. It is not ambiguous in the sense that the community is visualized but its boundaries are not clearly drawn. Curiously, several suggestions compete for this election of the nation. Bankim's language carries the mark of this indecision. It is rewarding to analyse these absent articulations—this combination of what it is possible to spell, what is not, of sight and blindness, of utterance and silence, to understand persistent difficulties in the formation of an Indian national identity.

An uncharacteristic awkwardness in the use of language indicates some ambiguity of this sort in Bankim. In essays we find the most diverse identities being called jātis. These range from castes (jāt or jāti) to Bengalis, to the Hindus, to Indians. Given this, it is in fact easy to determine the actual linguistic meaning of what this word means. If it is to carry clear meaning in all its places/syntactical positions and not collapse into a mass of contradictions, it means something generic and empty, like 'community'.

Ambiguity of the meaning of jāti in Bankim's essays therefore is an index of a historical difficulty of discourse; his trying to stretch its meaning in an unaccustomed direction; making it spell something which it was not accustomed to spelling out. This is why such a master of open and closed language, someone who is equally at home in the use of exaggeration and inflection, still runs out of his considerable resources, and says, external to the text, in a footnote standing in self-conscious isolation, that for the purposes of this essay, the term jāti means 'nationality' or 'nation'! Evidently, he is attempting a gerrymandering of the meaning-boundaries of this concept, cancelling its earlier indeterminacy, and fixing it on to the new, unfamiliar meaning.

Thus the term jāti in its new meaning of the nation remains interestingly empty. It is empty in the field of concepts of natural consciousness. It is external: European history teaches us what it is for a people to be a nation. However in Bengali it is not yet a descriptive term: on the contrary the other meanings of jāti—caste, religion, sect—are descriptive. They describe groups which exist and make sense to the natural consciousness. But in Bankim's use, it indicates a possible community which chooses itself as the collective actor for historical initiatives. The task then becomes one of naming the nation, electing it from among the many given contenders of caste, religion, language identities, or creating one that is larger, more complex, yet unnamed. Nation thus becomes the conferment of a title, and whoever becomes ultimately this 'nation' would, it is obvious even to Bankim, become entitled to be the privileged historical actor.

Bankim does not choose his nation easily, or without second thoughts. Of all others, this is the place in his thought which is most tense, most full of contradictory possibilities. He is among the few who can achieve a critical distance from the layered communities to which he belongs, and abstractly appreciate the need for a modern secular discourse: English education has provided him with that modified Archimedian point. Still, quite often, he inclines towards giving an answer in terms of a Hindu nation. The question of subjection is often asked in terms of a long seven centuries, though what to do with the residue if Hindus constitute the possible nation is never asked. Clearly, there are phases of his intellectual biography when this distinction is simply insignificant: the argument about the peasantry applies equally and irrespectively to Rahim Sheikh and Rāmā Kaivarta;[72] and apparently it is in their convergent economic interest to see the end of the colonial *zamindāri* structure. Tapan Raychaudhuri has recently suggested an intellectual rupture between a secular early period and a late Hindu one,[73] supported by Haraprasad Sastri's testimony.[74] Yet, it is difficult to accept that entirely, in view of the still earlier fragment of the poem on Akbar, which sees in the times of the Mughals both a literal and allegorical presence of subjection—a point made quite explicit in the poem's handling of temporality: its significant conflation of then with today.[75] In the absence of a theory of periods, it seems better to see this as a circuit of tension in his thought, a question whose urgency he obviously felt, but to which he could give only unclear and troubled answers.

Nowhere does his troubled mind show its lack of clarity more clearly than in his reflections on 'The independence and dependence of India'[76] (1879 according to J.C. Bagal). Modern Indians apparently believe that ancient India was free, and modern India dependent—to their serious regret. This requires, according to Bankim, two distinct but related enquiries: a theoretical enquiry about the meaning of terms like liberty and independence; and and historical comparison between the conditions of society in the periods of ancient freedom and modern bondage. The point of the comparison is to determine if people were happier in ancient India or under the present regime. 'By reading English Bengalis have learnt two words: liberty and independence . . . Many believe that the two words are synonymous. Common understanding has it that this means rule by people of one's own community. If the ruler is an alien, then his subjects are unfree, that kingdom is dependent. For this reason, India under the present British dispensation is called unfree and subject. For this, India under the Mughals or Bengal under Serajuddaula are called unfree and subject. Let us analyse the grounds of these beliefs.'[77]

The rule is too simple: because the Hanoverians were German and Napoleon hailed from Corsica. 'If we cannot call England under George I or Rome under Trojan colonised or subject, why would we do India under Shahjahan or Bengal under Alivardi?' Thus it appears that the simple fact that the ruler is of a different community does not turn a country into a colony.[78] Under the Mughals the earlier state lost its independence, but it did not lose its liberty. Bankim remarks subtly:

> In England there is a special use of [the terms] political liberty, but we are not obliged to use that meaning. Because that meaning is not relevant to the present analysis. We shall use it in the sense in which Indians understand it.
>
> If people from another country come to rule a land, this leads to an infliction. Those who are of the ruler's community enjoy greater eminence than the natives. This leads to oppression of the people by a foreign race. Cases in which there is such distinction between subjects of the native race and subjects of the ruling one we would call unfree. A country which is free of such racial oppression is free.
>
> Therefore, even countries which are under rulers from other parts can be free. For example, Hanover at the time of George I, or Kabul under the Mughals. Conversely, even independent kingdoms can be unfree, for instance, England under the Normans; or India under Aurangzeb. We regard India under Qutabuddin unfree and colonised; and India ruled by Akbar free and independent. However, ancient India was free and independent, modern India is unfree and colonised.[79]

Bankim's argument, remarkably, does not give any encouragement to usual forms of chauvinism. Ancient India did not have absentee foreign rulers, but arbitrary and authoritarian ones. Although there was no racial oppression in ancient times, it had an oppressive system of castes which was in some ways far worse. 'Some people may say, to compare English dominance with that of the Brahmins and Kshatriyas is inappropriate; for, Brahmins and Kshatriyas, even though oppressive towards Śūdras, were of the same race, the English are aliens. One wished to answer to this that to those who are oppressed, oppression by their own or aliens is similar. It does not appear that oppression by one's own tastes somewhat sweeter, and by others more bitter. But we do not wish to offer this answer. If someone has a taste for oppression by his own people, we do not object. Our only point is that in place of racial oppression in modern India there was caste oppression in ancient India. For most people, they are the same'.[80]

All his varying moves show on this point a strange combination of urgency and lack of decision. It is crucial to become a nation; but there are several descriptive identities already available to consciousness none of which seem to fit. Castes cannot become nations. Bengalis cannot become a nation, especially with their disappointing history. Hindus may become a nation, but they are not the only ones who live here, and they are irremediably divided against themselves, impossibly segmented for any possible collective action. None of the pre-existing identities can become a nation. The nation, then, has got to be invented.

The task of imagining a nation, however, has great implications for the project of history writing, conferring on this possible community a common welding history, a narrative of unity. Before this community is united by acts, it must be united by narrative. Obviously, there is a great internal problem of consistency; the narrative would show as already united a community which actually needs uniting. This is the basis of its temporal switch, putting narratively in the past what is believed to be possible in the future, in other words, simply showing the future as past. An imaginary community can only have an imaginary history. The actual history of Hindus and Indians could, by definition, never capture what was wanted of it, a history of united mobilized action. Only a fictional history can show such reconstructed Hindus or Indians, putting men of the future inside events of the past. That is why the task wanted of this historical discourse could never be accomplished by a discourse of facts, but by a discourse of truth, or poetry, of the imagination. This discourse thus moves, out of its internal logic, from the rational discourse of fact-gathering to the mythic discourse of the novels.

Already, within this rational discourse of facts, there are large cracks through which mythic construction shows its unmistakable form. One symbol is particularly significant. Despite all the confusion and second thoughts about the people who might constitute this magical 'we', there is a constant—the symbolic and literal territoriality, the great sign of the land, the country as the ground. Increasingly, in Bankim's discourse, the territorial identification gathers strength over the demographic one; the people may be problematic, but the land is certain, unquestionable, ever present. She turns into a feminine symbol of sustenance, the Mother. The *Bande Mātaram* is not statistically communal. Muslims provide, on any serious demographic enumeration, the majority of those twice seven crore uplifted hands, or the arithmetic of the song collapses. If the land is the mother, it becomes difficult to distinguish between her children. And a more consciously secular poet could say, years later, 'who dares ask: are they Hindus or Muslims? Helmsman, say they are men who are drowning, sons of my mother' (*Hindu nā orā muslim ei jignāse kon jan? kāṇḍāri, balo dubiche mānuṣ, santān mor mār*).[81]

Imaginary History

To enter into the last novels of Bankimchandra is to enter a peculiar time. It is a time constructed in imagination confused with the past. These novels are vital for an understanding of his mental world, because these complete in so far as such a thing could be completed what was begun in the myth of Kṛṣṇa. That was a myth of a single figure, a symbolic person of heroic action; here we see developing a wider myth of a people. Before the advent of rationalistic histories, history simply meant tales worth remembering and reciting. The purpose of such history is deeply Durkheimian—it provides for social cohesion, it gives the society of the present sometimes an entirely fictive 'memory'. Thus, what a society 'remembers' is not always what has actually happened; or at times, it believes in one of its many possible memories.

Curiously, Bankim himself said in his preface to *Rājsinha* that he had never written a historical novel before.[82] This must mean that in his own mind he was trying to do something quite different in these novels. In them we see gradually unfolding a landscape of imaginary effort against constituted historical injustice. And they enunciate, as Bankim himself claimed, a theory of anti-colonial *bildung* or *anuśīlan*. All these are novels of contestation of injustice, and persons becoming what they wish to become by intense willing and praxis.[83] However this depends on a successful coding in terms of time, and alerting his now devoted audience to its artifice. It is a simple code of a complicated time to be read off by the ideal reader through its device of false indexing. Ostensibly these events occur in the past, but they always point to their possibility in an imagined future. Without this, the theory of anuśīlan would misfire, creating, instead of optimistic praxis, a deep historical pessimism. Time is clearly linear, and the occurrence of glorious events is a result not of providence but of praxis.

A certain principle seems to govern Bankim's choice of points in history where these narratives are inserted. In a relation between history and historical fiction, there is a crucial position of some kind of underdetermination. In the novels Lukacs discusses, especially Scott, the novel usually selects a character whose features are underdetermined by known history, and this makes possible a fictional play of incidents, which weaves in and out of the known tracks of historical events.[84] The lives of the Darnays and the doctor's family has this relation with the known, and inside the novel, inexorable course of the French revolution.[85] Ivanhoe's fictive life similarly weaves in and out of the life story of Richard I.[86] At one level, Bankim's use of the historical frame is similar. He chose the less known, vague, factually underdetermined events from the historical record: the tussle of a lesser Rajput king with Aurangzeb, the historical rumours of Bhabani Pathak's robbery, the less unknown sannyāsī rebellion, and the little known broken chronicle of a small landlord in Bengal. Two structural devices are used to create this favourable historical climate for the novelistic story. First, of course is basic narrativization, the sovereign decision to choose the points of starting and ending the story. This is how the Rajput king can be shown undefeated by stopping the narrative at a point, while, if it continued, there was great danger of his or his successor's succumbing to Mughal pressure. A similar device is used in *Ānandamath*, where the story breaks off at a possible point of the rebel's victory, as opposed to the more melancholy ending in the historical record. Secondly, it is the factual sparseness of the record which is utilized, exploiting the softness of the evidence, so the counterfactuals of the fiction do not appear contrived. But the purpose of the fictional narrative is exactly the opposite of what Lukacs found in the European historical novel. These do not try to fill out, give a concrete sense of a historical period, heightening its historical truth, a point that Manzoni in his long essay made earlier.[87] The point of these novels is precisely to 'falsify' history. They try deliberately to probe and use counterfactuals, to extend those lines in the tree of eventuation which were not actually followed up, explore the peculiar terrain of history's nonactualized possibilities. They cannot, in most cases, falsify historical events themselves, but by placing fictional narratives in their midst they seek to shift and displace their meaning. One of the main techniques in this is the creation of this double-valued time, a mixture of the past and the future, a time that is grammatically indescribable.

There is a clear line of development from *Kamalākānta* towards both the *Kṛṣṇacaritra* and the later novels. No reading of *Kamalākānta* is complete if it reads merely its sparkling humour and not the dark anguish which forms its background. Bankim is, to perpetrate a pun on Barthes, the greatest unhappy writer, who does not forget to remind us of our duty, responsibility, until the world is made better, of unhappiness.[88] Often the link between these passages in *Kamalākānta* and those of the later novels is quite literal. It is interesting to find that the famous scene in the underground temple in *Ānandamath*, in which not just Mahendra but his countrymen are first shown the identification of the mother and the land, is not begun inside that novel but in *Kamalākānta*.

> Why did I take this overdose of opium? Why did I take opium? Why did I go to have a look at the idol? *Why did I see what I never shall*? Who showed me this magic?
>
> This is what I saw: suddenly I was floating helplessly on the great stream of time flooding the horizon, I saw—in the eternal, endless dark great waves lashed by a great storm—occasionally luminous stars rise and go out and rise again. I was entirely alone. I felt afraid of my loneliness, entirely lonely, without a mother, calling for my mother. I had come to this sea of time searching for my mother. Where is the mother? Where is my mother? Where is Bengal who had given life to Kamalākānta? Where are you in this terrible ocean of time? Suddenly a heavenly music filled my ears, a bright red light like the rising sun broke out in the horizon, a gentle breeze started blowing—and I saw in a great distance over these tumultuous waters draped in gold an idol of the *saptami* day. It was laughing, floating on the water, radiating illumination. Is this my mother? Yes, it was her. I recognised her, this was my mother, my country, of this land, in the form of this land, decked with infinite riches, but now submerged in the waters of time. Ten arms decked with jewels were stretched in ten directions, carrying various powers as different weapons, the enemy destroyed at her feet, the powerful lion under her demolishing the enemy. I would not see this form now, not today, nor tomorrow, not before the deluge of time is passed; but we will see it some day . . . Resplendent in new glory, carrier of new strength, filled with a new pride, seeing new dreams . . .[89]

There is an almost identical passage in *Ānandamath*, which I shall discuss later. And what is remarkable, but often less noticed, is a movement of a new language of sentiment, of poetry even here, although Bankim was an astute enough observer of language to realize that his poetry was bad, as indeed it remained in the *Bande Mātaram*.[90] However, the point is not that the poetry is inexact, but that it is poetry. And there is the same, remarkably mixed language in this poem,[91] which I consider a lineal ancestor of *Bande Mātaram*. Sanskrit provides the initial moves creating an atmosphere of sombreness and grandeur, and the main cast into which sentiments are poured; but the breakdown, or if you like, the quick transcendence of that formal Sanskrit is equally significant. Use of Sanskrit is obligatory, because it is the mandatory vehicle of classicization; but its correct mould is breached—to the great annoyance of purists—by a more intense, spontaneous Bengali, though remarkably the proportion of Bengali to Sanskrit in this poem is much less. Sanskrit purists misjudged the question of linguistic correctness, did not realize that the question of 'correctness' was wholly displaced here—a question of not being grammatically but politically correct. Indeed, the ungrammaticality was in a sense perhaps constitutive of the intention behind the poem, because it set itself a contrary, conventionally impossible task. It was to classicize and popularize an idea at the same time, thus it must not be something so arcane as to be meaningless to ordinary people. Ordinary people customarily used stotras after all, and it would be wrong to state that they did so entirely uncomprehendingly. In trying to achieve this contradictory speech—Sanskrit and the vernacular, grand and spontaneous at the same time—Bankim transgresses all linguistic norms; and under the pressure of rising sentiment the language breaks into vernacular and a kind of low Sanskrit accessible even to common people.

But since I see the anticipation of *Bande Mātaram* in this poem, some comparisons between the two would be in order. The *Kamalākānta* song is evidently the prefigure, the seed of *Bande Mātaram*, but quite different on some points. Its movement is, internally, from lighter to heavier Sanskrit, and from general terms to technical doctrinal identifications from the twenty-second line, where it merges with a quote from the *āryāstotra*. In *Bande Mātaram* the internal movement is in the reverse direction—from clear Sanskrit to even clearer vernacular. Indeed what is grammatically a false step is clearly a step of genius in the later song—it is the point of symbolic contact between the classical and the popular. That is Bankim's popular slip, giving us a glimpse of his 'political unconscious'. *Bande Mātaram* accomplishes the most extraordinary transformation, an unprecedented mixture of discourses, moving quickly and magisterially from the form of the stotra to the form of a slogan, from a private incantation to the public, collective chant of a people in movement. Finally, the two stotras derive their meanings from the immediate discourse which frames them, in *Ānandamath*, of an ascending seriousness, in *Kamalākānta* an ineradicable lightness of speaking, and these are inscribed in the poems. The *Kamalākānta* stotra still bears the inimitable, unmistakable comic signature—the frivolous line: *jay mā Kamalākāntapālike* (glory to the sustainer of Kamalākānta). What a peculiar conservative, who dares indulge in facetiousness with the divine mother herself! In *Ānandamath* this humour, this possibility of double meaning is erased; for it is of a deadly seriousness that it is possible to read only one meaning in that song. The attributes of the goddess are significant: like the discourse of reconstructed Hinduism, she is new, reinvented, a political goddess. Even the language displays the tense instability, experimentalism, the newness of what it says—a string of suspect *samāsas—navarāgaranginī, navabaladhārinī navadarpe-darpinī, navaswapnadarśinī*. It is unclear whether she sees these new dreams herself or makes others, her seven crore progeny, see them. But what does it matter? She is a Durkheimian goddess, a sacred name of the collective self. Her seeing these dreams and making her children see them are the same. Thus the poem fashions an ungrammatic phraseology of poetic perfection, adequate for the sentiment building up in them and seeking expression.

From my angle the essay *Āmār Durgotsav* provides the critical connection. Of all the pieces in *Kamalākānta* this is the least funny. It already presages a new enthusiastic style of writing, a Bengali style that has appropriated the repetitive, intensifying manner of the use of adjectives in Sanskrit, a style that allows Bengali to appropriate the stotra form, and break it down into something that can be easily spoken and included in prose. The enormous distance in tone and signification between the two songs is above all by a single word, a single discordant sentence. That single word is the seemingly innocent, apparently unremarked *āmār* (mine). In the first stotra the figure of Durgā is invoked by the standard chant—*sarvamangalamangalye, śive āmār sarvārthasādhike* (notice the shift of the *stava* into unmetric Bengali prose): but that single interposed word *āmār* disrupts the identity of the *stava*; it does not remain what it was. I consider this irreverent play crucial to Bankim's personal thought—because it establishes a right to put traditional forms to wholly unconventional use. Thus the stava can become a matter of fun and humour, just as the playful, erotic Kṛṣṇa can be turned into the unaccustomed centre of a political ideology. When, in his second movement, Bankim suppresses that *āmār* and comes back to the stava in its serious form, this does not index a movement from an ironical to a traditional stava, rather a movement on the same plane from an ironical individual salvation to a grimly serious collective one.

A similar point is made by inserting the line *jay mā Kamalākāntapālike*, and later its absence. Kamalākānta's stance—of being obliquely related to the world, of a delightful, pleasurable wrongsidedness (*besuro bujhi, besuro śuni, besuro likhi*) is shown as an unstable equipoise which was meant to be disrupted. In an inverted world, Kamalākānta with his gift of truth, naturally had to see things upside down. But the next logical step is to try to invert the inverted world, and the preliminary skirmishes with colonialism, though a battle in discourse, is begun by him in the court scene. To continue that it requires a larger, more tragic, less verbal theatre, which the novels provide by creating the great theatre of war. The transformation of the jester has begun, he is already turning into the priest.

Construction of the Icon

In the conventional historiography of nationalism, Bankimchandra's fame, and to westernist secularists his ignominy, lies primarily in his responsibility for dreaming up the icon of the nation—feminine, maternal, territorial, natural, but offensively Hindu. To ahistorical secularists, this gives him a fair share of the responsibility for the Partition and some for the frequent communal riots of the present. However, as Tapan Raychaudhuri has shown recently, although these symbols are taken from the repertoire of Hinduism, they are actually highly unorthodox. Since this icon is constructed most effectively in *Ānandamath* let us devote some attention to exactly how this is done. The song *Bande Mātaram* is first sung by Bhavānanda. After the first three lines, Mahendra, who is walking by his side, is confused.

> Mahendra was somewhat astonished by the song, could not understand it—this mother endowed with water, fruits, cooled by spring winds, verdant with crop, who is this mother? 'Who is Mother?' he asked.
>
> Without replying, Bhavānanda went on singing,
>
> śubhra jyotsnā pulakita yāminīm,
> phullakusumita drumadalaśobhinīm
>
> suhāsinim sumadhurabhāṣinīm
> sukhadām varadām mātaram.
>
> Mahendra said: 'But this is my country; this is not my mother!'
> Bhavānanda said: 'We do not recognise any mother besides her—
> *jananī janmabhūmiśca swargādapi garīyasī* (one's mother and one's motherland are nobler than the heavens). We say, our motherland is the mother, we have no mother, father, brothers, friends, wives, sons, houses, families, we have only her who is endowed with water and fruits, cooled by the winds of the spring, green with corn . . .
> Then Mahendra understood and said, 'Then sing once more'.[92]

The dialogue between Bhavānanda and Mahendra is clearly a transposed dialogue between the author and his audience, each of Bankim's readers, because to see the country as the mother is highly non-standard in Hindu mythology. It is a transposition, using traditional figures certainly, but to purposes that are highly innovative.

The most obvious sense in which this is untraditional is the political message. One of the most curious elements in the historical career of the Hindu religion had been its unwillingness to be turned into a political creed, and its apparently happy acceptance of all political authority as a marginal but necessarily evil aspect of social life.

Mahendra is ushered into an underground temple, and presented to an idol which is remarkably different from any that practising Hindus could have actually worshipped.

> . . . slowly he could see a monumental image of four arms (*caturbhuja mūrti*), carrying *sankha, cakra, gadā* and *padma*, with the emerald *kaustubha* adorning the chest; the wheel *sudarśana* placed as though spinning in front. Two large severed heads, blood-spattered, placed in front representing Madhu and Kaitabha. To the left stood Lakṣmī, with her hair let down, with a garland of lotuses, but apparently frightened. To the right, Sarasvatī, surrounded by the book, musical instruments and figures of the *rāgas* and *rāginīs*. On the lap of Viṣṇu was a marvellous form—she was more beautiful than Lakṣmī and Sarasvatī, with more grandeur than either. The *gandharvas, kinnaras*, gods, *yakṣas* and demons were worshipping her. The hermit spoke to Mahendra in a very solemn, very awed voice: 'Can you see everything?' 'Yes,' said Mahendra.
>
> H. Have you seen the figure on the lap of Viṣṇu?
> M. I have; who is she?
> H. Mother.
> M. Who is Mother?
> H. She is the mother whose children we all are.
> M. Who is she?
> H. You will recognise her when the time comes for you. Now, say, Hail Mother (*Bande Mātaram*). Come along and watch.
>
> Then the hermit led Mahendra into a second chamber. There Mahendra saw a wonderful figure, of Jagaddhātrī (Mother of the Universe), rich in every limb, decorated with every possible ornament. Mahendra asked: Who is she?
> H. What our Mother used to be.
> M. What is that?
> H. She had tamed the animals of the forests, the elephants and the lions, and in their habitat she had established her own abode. She was adorned by all ornaments, laughing, beautiful. She was gifted with all the riches of the earth, the colour of the rising sun, Bow to her.
> Mahendra bowed with great emotion to the motherland in the form of the Jagaddhātrī, and the hermit showed him a dark tunnel and said, 'Come this way.' The hermit led the way, Mahendra followed him in trepidation. In a chamber deep under the ground, there was some strange source of meagre light. In the faint light he saw an image of Kālī.
> The hermit said, 'Look, what our Mother has become.'
> Mahendra whispered in fear, 'Kālī.'
> H. Kālī—covered by darkness, of a black colour. She has been robbed of everything, that is why she is nude. Today the whole country is a graveyard, that is why our Mother has a garland of bones, dead skulls. She is trampling her own welfare under her feet. Alas!

Tears streamed down the hermit's eyes. Mahendra asked why does she a have sword and scythe in her hands?

H. We are her children, we have given her these weapons. Say, *Bande Mātaram*. Mahendra said *Bande Mātaram* and bowed to the Kālī idol. The hermit said, 'come this way', and started climbing a second tunnel. Suddenly, their eyes were dazzled by the rays of the morning sun. Sweet singing of birds sounded on all sides. They saw a ten-armed golden idol in the middle of a marble temple, laughing in the light of the morning sun. The hermit bowed to her and said, This is what our Mother will become. Ten arms are spread in ten directions; in each a different weapon, a symbol of a different power, her enemy sprawled defeated at her feet, the lion she rides on mauling her adversary. Arms spread in different directions . . . '

Satyananda was overpowered by emotion and spoke through tears. 'Arms spread in all directions, carrying various weapons, overpowering enemies, rising on the lion, fate as Lakṣmī on the right, on the left Sarasvatī the giver of knowledge and science, accompanied by Kārtikeya, the symbol of strength, and Gaṇeśa, symbol of success. Come let us both bow to the Mother. Then both of them looked upwards, and sang with folded hands,

> *Sarvamangalamangalye śive sarvārthasādhike*
> *śaraṇye tryamvake gauri nārayanī namohstute.*

When they got up after bowing to her with great devotion, Mahendra asked with emotion, 'When shall we see our Mother in this form?'

The hermit answered, 'The day all her children would call her their mother, she will be pleased with us.'[93]

Several dislocations are discernible in this text, which would be of interest to us to measure.

Let us take up the questions internal to Bankim's texts first. In contrast to the jocular tone of the *Kamalākānta* stotras which take surprising liberties with tradition, these return to a more serious tone. This is not, however, a return to an unaltered past point. The *Kamalākānta stotras* could be written only by someone in a strangely ironical proximity with Hindu ritualism, someone who is wholly at home in them, because his way of parodying, perhaps all parody, requires a great adeptness. Parody defiles by the slightest of deflections, not by a frontal attack; and thus needs to be done by a cultural fifth columnist. This piece parodies the *stotra* form, but at the same time points to the private and lonely, non-collective nature of the *mokṣa*, the deliverance sought. The traditional Hindu would of course abjectly debase himself to his deity to seek mokṣa for himself. Sometimes, he might include in his application for spiritual preferment his immediate family; but there are philosophical arguments which suggest that such journeys are best undertaken alone. *Kā tava kāntā, kaste putraḥ?* Who is your wife, or your son, to claim a share of your salvation? So the

Hindu supplication to his deity can be made as part of a community, but it is an intensely lonely prayer. *Kamalākānta* parodies its intense selfishness: the glory of the mother is seen to lie not in benevolence towards the world in general, but having a special and declared bias in favour of Kamalākānta's desires (āmār sarvārthasādhike). Selfish accumulativeness of this kind is not at all uncommon in Hindu prayers.

The parodic use displaces the stotra from its traditional appointed place in ritual schema, and makes it into something that is available in a new way, a form which could be turned to new, unprecedented use. It is because of the earlier parody that the later political appropriation is possible.

What is truly remarkable is the great innovativeness of this free form and its new use. Earlier, the devotee, though sitting in a crowded temple was essentially alone, praying for private and individual favours. The temple crowd was therefore a devout sack of potatoes. In Bankim's new invocation, the utterance is collective. It is not just an accidental use of the plural verb, it is a theoretical use which underlines that if not enunciated by a plurality, by all who are figuratively represented inside the song, it is in a sense meaningless.[94] Apart from the collective pronunciation, the community becomes the object of an almost classically Durkheimian self-adoration, except that the indirectness, the maskedness of the object is dispensed with. The community, the children of the mother, are initially invoking her; but the more her attributes are enumerated, she becomes the same community under a different, ideal description. Initially, she is represented as the land, *sujalā*, *suphalā*, *malayajaśitalā*, *śasyaśyāmalā*. But the main point of the song is for this figure to act, and it is the people, its inhabitants, who can bring about such action. The mother's attributes gradually become collectively anthropomorphic; *saptakotikanthakalakalaninādakarāla*, and her features all become action features; *bahuvaladhārinī*, *tārinī*, *ripudalabārinī*. At this point, the elaborate form of the mother, the elaborate ritual, is simply a mediation of the collective self to itself.

It would be false in this sense to see the icon that is created as a conventional or traditional Hindu icon. Its attributes do of course come from the traditional discourse of Hinduism. Nature is often seen as a sustainer and therefore deserving worship, from the Upaniṣadic hymns to natural deities to the more mundane and philosophic doctrine of the aesthetic availability of the world in bhakti movements. Still, this is something similar but fundamentally different. The bounteous nature of earlier religion was not appealed to for large avenging actions. Also the distinctness of human beings from nature as the object of worship was clear in those earlier forms. In the new religion it is not only possible to worship, offer prayers, sing the glory of the self in a mediated form; this mediation, indirectness, masking of the collectivity through its own metaphor must be seen through, and dissolved: thus the worship should, to be successful, move historically from a prayer to the metaphorical to the literal self. The destiny of this doctrine would be fulfilled only when, half surely at first, but with increasing passion and decision, the community comes to treat itself as eternal and offers itself its own prayers. It thus transfers to a modern identity a traditional form of emotionalism.

I wish to claim that far from being a move that was possible within traditional religious discourse, this is possible only on condition of a certain unprincipled use of Hinduism. Yet, in the act of doing this, there is an implicit and decisive recognition of the nature of the elasticity of a tradition, of the fact that a tradition has such great longevity precisely because it is internally diverse and it is possible to stretch its resources to make it accomplish tasks it had never undertaken before. A tradition of eternal repetition cannot survive historical change. It is significant that Bankim was not among those of his generation who stayed within the received Hindu tradition, and attempted to fashion intellectual defences in its favour. He was a person who returned to a position that was of a reconstructed Hinduism very different from its traditional form, and from that point of view, rightly regarded with suspicion by those more devoutly inclined. On at least two points, this reconstructed philosophy diverged sharply from its conventional form. The apparatus of argument it used was almost wholly rationalistic, and it was prepared, a little too easily, to surrender as nonsensical a great deal of the deeply held certainties of traditional religion.[95] A traditionalist could therefore ask what was historically Hindu in his construction of philosophic and ethical doctrine. A vaiṣṇava could object on similar lines to a vaiṣnavism which so casually dismissed Caitanya.[96] Bankim's answer, stated explicitly, was that this was the only way Hinduism could survive in history.[97] Secondly, there was in this reconstruction a great deal of personal idiosyncrasy with which more traditional believers would be out of sympathy.

In one of his passionate sequences the guru in *Dharmatattva* tells his pupil:

> This *vaiṣṇava dharma* recounted by Prahlada is the finest of all religions. This is the essence of religion, and is therefore present in all pure religions. To the extent to which a religion is pure, it subsists in it. The Christian and Brahmo religions are a part of this *vaiṣṇava* religion. Whether I call god, or I call *allah*, I address that Viṣṇu, the lord of the universe. The person who has come to realise the consciousness full of knowledge and joy which is the soul of all things; he who considers all being as his self; he who is above all distinctions, or a person who strives to attain a state of mind of that kind alone is a *vaiṣṇava* and a Hindu. He who constantly shows malice towards others, does harm to others, quarrels with others, who is engaged in ostracising people for deviating from caste practices, I would not call him a Hindu even though he may have a whole bunch of sacred threads on his neck, a mark covering his

entire forehead, a pigtail on his head, a *nāmāvalī* draped around his body and constant incantation of Hari on his lips. He is a *mlecha* worse than all *mlecchas*; a Hindu loses his Hinduism even by contact with him.[98]

In one of his direct expositions of this reconstructed Hinduism, Bankim was still more explicit about his heterodoxies. He declares that 'it is my firm conviction that without a revival of the religion of the nation there is no good possible for India'. But the first question is what is the Hindu religion?

We find many types of Hinduism. A Hindu does not go on a journey if someone sneezes; says "true, true" if a lizard squeaks; snaps his fingers if he yawns. Is all this Hinduism? You should not sleep with the head in that direction, or eat facing the other, not go out after seeing an empty pot, not shave on a particular day, desist from some other work on another. Is all this the Hindu's religion? Many would admit that such things are not Hinduism, simply the rituals of the stupid. If this is Hinduism, I should like to declare clearly that I do not want its resurrection . . .

I know a zamindar, a Brahmin by caste and punctiliously Hindu. Summer or winter, every day, he would get up very early and take his bath. He would begin his worship right then and continue with them till late in the morning. He considers it a great misfortune if there is any hindrance to his practice of ritual observances. In early afternoon, he sits down to a frugal, strictly vegetarian meal. After lunch he takes on the work of his zamindari. At that time he concentrates on which *ryots* he would ruin, which helpless widow's property he might confiscate, whose loan he would leave unpaid, who has to be sent to the jail by forging documents, for which case false witnesses have to be arranged. We happen to know that the regard this gentleman has for religious rituals, gods and Brahmins is entirely genuine; there is no element of pretension there. He incants Hari's name even in the act of forging documents; indeed, he believes that if he remembers Hari at this moment, his forgery must bear fruit. Is this man a Hindu?

Let me speak of another Hindu. There is nothing that he does not eat; he takes everything except what is bad for health. And though a Brahmin he is a moderate drinker. He takes food from all castes; does not refuse dining with Muslims and Christians. He does not observe a single ritual of worship. But he never tells lies . . . Unselfishly he gives to others and tries to do good to them. He controls his senses as far as he can, and honours God inside himself. He never deceives anyone, does not covet other's goods. He considers his senses and objects like the sky to be expressions of God and of his power and beauty, and thus pays them mental homage. Since he considers the image of Krishna as recounted in the *Puranas* the complete expression of all-powerful God, he calls himself a *vaiṣṇava*. According to the precepts of Hindu religion, he treats his elders with regard, maintains his family with love and care, is kind to animals. He is above anger, and full of forgiveness. Is this man a Hindu? Of these two who is a Hindu in reality? Or is neither? If neither, what are they? If we do not get Hinduism in either of them, then what does Hinduism consist in? One has strayed from *dharma*, the other from religious rituals. Are the rituals religion, or is *dharma* religion? If rituals are not religion, but *dharma* is, then we have to call that man who has strayed from the rituals the truly religious person. What are the difficulties in saying this?[99]

The rationalist who has returned to Hinduism cannot return to a religion which is identical to the one he had rejected.

If there is indeed a world inside fiction created by an inversion of the referential power of words, the world of Bankim's last novels is quite strange. It is a world of wars. If it could be turned visual, this would present an endless scene of armies, moving people, clashes of great collective violence, a world from over which the smoke of battle never wafts away. This is not simply because it is a world of the past. Bankim could have painted pictures of past domesticity, idylls of quotidian happiness (or misery) from the depths of the middle ages. The division is more fundamental than a simple one of time. The battles or conflicts never cease to allow much time for a clearer glimpse of the private destiny of the principal actors. Sometimes, there is a sort of two-plane story; a foregrounded story of individuals seen in the noisy, confused background of conflicts between large forces seen at a distance.

Common to all these stories are some formal features. First, in all these narratives, there is a symmetry between the occasions of these wars. Historically, all of them are hung on a wispy, insecure footing, some half-sentence of resistance, an inconspicuous footnote mentioning defiance. Small, practically undecipherable incidents are magnified into the major stories of these novels. For his purposes Bankim needs incidents which are factually underdetermined.

Inside these stories themselves there is a peculiar disequilibrium. The oppressors are historical, the rebels are fictionally created. All these are cases where there was a true, constituted *state* of injustice, or a ruling power which face (in the world of his fiction) a surprisingly small but invincible army of justice fighting a just war. Of course they are invincible because the narrator is their ally. A significant second feature is the emergence of a new actor—a *collective* actor in each case, a movement, an incipient people. Surely there is still a play on private and public destinies, of what happens to the individual and to the collectivity, but the balance between them is inverted. The collective becomes more powerful, and comes to the front. In all these stories, there is an armed people who avenge themselves. On the surface, Hemcandra's story in *Mṛṇālinī* is also an avenging story, but it is still a private battle. Jagatsinha does head a real army; but it is not a popular army, an army of justice.

This is a strikingly unusual picture for contents of Bengali novels, produced and consumed by the educated comfortable middle-class enjoying the colonial peace of the late nineteenth century. Novels in the

Bengali language were typically and standardly about romances, loves and heartbreaks, intrigues, ideas, at worst, business failures or bankruptcies of landlord estates. Its setting is basically interior—inside homes and gardens, city streets, village lanes, rural wells. In it, an educated woman taking an unaccompanied journey on a crowded steamer is considered the ultimate in risk-taking in a hostile world.[100] It is in a world of such small dimensions, in private spaces, that Bengali middle-class lives were played out, and novels faithfully reflected them. Bankim's world is quite different. Where does this world come from?

Although there are peculiarities of narratives in each novel, I see them as making a powerfully stated single argument about subjection and history, about the relation between leaders and the people and above all between the past and the future. Bankim himself suggests at various points a unity of all his later work—*Kṛṣṇacaritra*, *Dharmatattva*, the unfinished commentary on the *Gītā*, *Ānandamath*, *Devī Caudhurānī and Sitārām*; I stretch this and include *Rājsinha*, its last and practically rewritten edition.[101] We have seen earlier how Bankim invokes the strongly mythicizing, non-rationalist conception of history from the Indian tradition:

> *dharmārthakāmamokṣānāmupadeśa samanvitam*
> *purāvṛttakathāyuktam itihāsam pracakṣyate*[102]

In this already fact-related criteria are relegated to the background and the imaginative, mythical element of history brought out. According to this view, the primary purpose of history is to instruct, not inform; but it can obviously instruct in two ways. It can instruct in its factualist form, because it can tell us, through analysis, why what had happened happen. History however was not just that theatre of what has occurred, but also of what will—the undetermined future. Societies, he had established in the *Kṛṣṇacaritra*, give coherence to them through stories they tell about themselves, because people can gradually become what they believe they are. Time in history is an unending possibility for doing so; thus, this becoming of the self 'can happen any time'. Remark the oddness of the phrase—its gawkiness and force—*bāngālir erūp mānasik abasthā je kono samaye haite pāre.*[103] That could be done only by an imaginary theatre of conflated past and future and making men and women act there: in other words, history simply cannot perform its function unless it is dual—in part what happened, in part what did not, but could in the past, to be replayed in the future—in this charged, peculiar, beautiful time that defied characterization as tense.

In many works of this generation, we find an absent, unpronounced centre, and the parts are subject to a clearly readable *décalage*. Questions formed in *Kamalākānta* are given answers in *Rājsinha*. Strange connections bind the *Economic History* to the amorous adventures of adolescent Rajputs and Marathas. Who in his senses can see the *Economic History of India* being supported by two novels? But who in his interpretative sense, can indeed miss this?

Imagined communities like nations are of course fictions in a sense; but the process by which this fiction is created is real and not unamenable to critical examination. Bankim's fictive imaginings about his nation are of special interest for two different but mutually reinforcing reasons. Undoubtedly, he was one of the first in India to move in this direction: in order to fill the evident inadequacies of the Bengali character he began to look for and construct a more complex and much larger 'we'. Since others in other regions were driven by parallel self-doubts and agony, they were driven to similar devices. Secondly, precisely because Bankim is one of the first to undertake such imaginings, its logic—both the pressure which drives it forward and the anxiety which accompanies it—is particularly apparent in his works.

What is significant in *Rājsinha* is hidden on its surface—not the plot, but the choice of the theme itself. Formally, it is quite different from the hazy familiarity with Rajput warriors in *Durgeśnandinī* or in *Mṛṇālinī*.[104] As, because of accumulating evidence, thinking a glorious history of the Bengali 'we' becomes impossible, the boundaries of this 'we' are gerrymandered, and the 'we' opens outwards to include those who would have been conclusively regarded as 'them' by an earlier mode of thinking. Todd's annals and antiquities of Rajasthan add a wholly new dimension to the Bengali imagination, and decisively alter the structure of the inner world of literature. Bengalis gradually appropriate this 'glorious' history with eagerness, and begin to see, imaginatively of course, this history of distant, unrelated people as a sort of 'prehistory' of themselves. How many alterations this required, for this supposition to work! Evidently, this connection had no warrant in history in a narrower sense: it could not be a causal or even a lineal connection—i.e., the chronicle, lineally—of people living in the same geographical area and, in that sense, the 'same' people. Such connections can only be imaginative or artistic. But the problem of 'truth' involved in them is subtle. Quite often the historical narrative involved in these stories is not simply false. The history of the Rajputs and the Marathas or the Sikhs may indeed be true;[105] what is not true in quite that straightforward sense is the assumed idea that these are 'natural' ancestors of the modern Bengali. This becomes plausible through the growth of a clearly different 'we'. This 'we' is both a fiction and a reality. Assuming that it is a reality in the present, it was still a fiction when entrenched into the past.

Using its ineradicable advantages, the present always gerrymanders the past's frontiers and identities. This is not unusual. It is a constant feature of all societies, though the form of such story-telling may differ. Myths, legends, literatures, chronicles, novels, poems, anthems all tell these most socially urgent of tales. History telling is its most recent, and for the last two hundred years, most acceptable form. But in the excitement with such historical stories, in the celebration of their truth, analysts are apt to forget a most significant aspect of these. This 'themselves' of which the stories are told, is constructed partly by these stories themselves. Presumably, Rajputs in the sixteenth century would have been surprised to be told that their exploits would delight future Bengalis; just as Bengali contemporaries would have been to learn that later Rajputs may be proud of whatever these early Bengalis may have done with distinction. Both Rajputs and Bengalis would have simply considered each other as distant, unconnected people. Had these past Rajputs or Bengalis written their 'own' histories (and as Bankim never wearied of saying, no one else writes histories) it is highly unlikely that they would have included the others' accomplishments in their narrative. Today, not only these but many other groups routinely do so, and it is regarded quite proper for any 'Indian' to feel proud of the achievements of such a composite past. These histories are therefore really, tangibly, changed. Events of a locality become the events of the history of a larger nation, and in that sense, they do not remain the same events any longer.

Rājsinha begins this play with time. On many levels, this is a story of a new type compared to his earlier novels with a historical background. In *Rājsinha* the story moves in terms of space outside Bengal, and becomes in a more real sense a historic one. This does not mean historical factual authenticity: for its exulting end is due to the closed nature of a narrative. Within the fictive world of the novel, Rajput power is not humiliated;[106] but it is in real historical record. *Rājsinha*, then, shows a number of features which are not attributes of the historical figure. He is partly a mythic figure, partly a reconstructed babu, a symbol in whom several clusters of features mix and several models intersect at a critical node. He becomes, in precisely Barthes' sense a 'mythic symbol', partly babu, partly warrior, partly created by the novelist's dream of an alternate self. He is not the petty chieftain of an obscure segment of Rajput history; he is a figure coming to us out of the future. It is not surprising that many of these heroic figures, in their moments of comparative idleness from heroic action, behave in a way extraordinarily similar to the manners of nineteenth-century Calcuttans.

This reflects, in my view, the founding movement of conceiving a 'national' community, the historic beginning of an imaginative integration. The process is no doubt heavy with ironies, and it is a mixture of trends both likeable and unworthy. From one point of view, it simply shows the confidence of the educated Bengali's chauvinism—a process in which the Bengali aggressively appropriates the other. Bengalis do not as yet see themselves as part of a larger whole; they simply append India to themselves. They go out on a great imaginative journey across the subcontinent opportunistically selecting episodes from other people's histories and adding them entirely without reason to their own. At the same time, this implies a faint, uncharacteristic admission of inadequacy, of being unable to cope with British colonialism singlehanded. A peculiar alchemy of misunderstandings and misperceptions sets in. By looking for the attributes missing in the Bengali babu he begins shifting the boundary of his 'we'. Ultimately, it is through this process, reciprocally carried out from all groups, that the imaginary community of the nation gets formed. Also, this explains the peculiar fascination of the sedentary Bengali middle class for the horse-riding, athletic, extrovert Rajput (this may not be the historical Rajput, but that matters little really); why he so dominates the Bengali fictive imagination. The Bengali cannot shake him off, because he is his other, what the Bengali knows he needs, but what he equally knows he lacks.

However, there was a parallel search for heroic episodes inside the history of Bengal, and the next two stories in Bankim came from such romantic legends. These are great dramas of anuśilan, of believing something about oneself, and becoming that through great acts of the will. But this willed transformation is not of the individual, but of a mobilized collectivity, of the leader and the people. *Within the narrative*, in all these novels, the war ends in a victory of the army of the people—of the armies of Rājsinha, Satyānanda, Devī and Sitārām, led by their popular leaders. After victory, however, in each case, something inexplicable happens. Victorious armies simply melt away; their leaders turn to an uncharacteristic domesticity or exile—both anyway are exiles of different kinds. The main protagonists do this on the weak excuse that 'the time' has not yet come. They are told by higher voices, thinly disguised voices of history, that the British would rule for yet a long time to come. But the difficulty with this literal reading is evident. Although these prophets confer certificates on the English for their just rule, after *Kamalākānta*, without forgetting everything he showed through his sarcastic vision, the audience may have doubts about this judgement. Within these announcements themselves, acquiescence is made clearly conditional; and if these conditions are not observed, the historical situation could be regarded as open once more, where the contract of political obligation can be renegotiated.[107]

Bankim's narrational technique clearly used counterfactuals, which, while marking the fictive off from the actual, kept them in logical contact. It was a dream history of India, but more meaningful than actual history. Indeed, through the misleadingly religious looking doctrine of bhakti he had suggested a theory of mobilization. Clearly, in the last novels there is a suggestion about bringing together an organized people and a directing element through their common allegiance to an ideal, a common object of political action. So, the willingness of the common people to be directed is not based on a feudal principle of subjection to leaders of higher rank, but a recreated principle of exemplary allegiance to an ideal commonly held. Note some indices in the novels of anuśilan. Nearly all who become leaders are actors who are outsiders; they are not feudal status holders who turn against external oppression. Indeed, these indices seem to change fundamentally in the last novels. Earlier, Hemcandra and Mīr Kāssim were status-holding protagonists. They fail, because Hemcandra misreads the task of resistance as that of an individual warrior. Mīr Kāssim relies on old traditional ties of loyalty—mercenary or patronic. Hemcandra has no army; Mīr Kāssim's army melts away at the time of crisis against superior European organization and sense of purpose. Resistance against oppression and unjust external power requires, within Bankim's fictional world too, a different solution.

If this reading is right, then, despite the fact that they depict different times and spaces, the battlefields are, as a sign, identical. It is through this identical metaphorical field of resistance that Hemcandra, Mīr Kāssim, Pratāp, Rājsinha, Satyānanda, Devī and Sitārām undertake their individual journeys of violence. The results are vastly different; all believe in violent retribution, but they organize violence differently. Let us briefly look at their principles of organization.

Except for Rājsinha the last three do not depend on rank; they are leaders risen from the people. So the right of the leader to the unquestioned loyalty of followers is not based on primordial loyalty of any sort; it is a right accruing from their ability to represent in an exemplary fashion what the ordinary members of the movements also cherish. Thus comes into play a triangle of leader, people and god, the geometry of a successful ideology of resistance. It implies that the leadership is not an attribute of the person, but of a structure of relationship. The leader–people structure is ruptured the moment the leader leaves his vocation, as Sitārām does. He misreads his own status, and tries, in the period of his narrative decline, to translate his populist kingship into a titular conventional one. Actually, he is not a conventional king, but a leader of his people; and this translation consequently fails. Only when he wishes to resume his populist title is he, for a tragic moment, back into an active position. It is more compatible with traditional thought to see the ordinary mortal as following, half-worshipping a leader, avatara, or seer, who in turn is a follower and reflection of God. The leader is to them what God is to the avatara. Hierarchies of approximation and emulation are fairly central to Hindu religious thinking. Bankim however is suggesting something far more modern, which is captured by a triangle of the devotee, the leader and the deity.

Other figures and their destinies too confirm this narrative logic. Rājsinha was not the only feudal ruler in Rajputana, but he goes through an election by history, because he alone responds to the challenge of collective ignominy. Like Sitārām, he is able to defeat the constituted army of injustice because he is willing to be transformed from a mere royal figure to a leader of his people. One cannot, thus, inherit leadership of just battles, one has to be elected, undergo an appointment by history. Such popular loyalties, won by organizing just resistance, articulating the sense of injustice of a people, cannot be maintained except in a theatre of action, a continuing popular movement. These leaders come from indistinct, at any rate non-aristocratic backgrounds, with nothing extraordinary in their previous lives; they are moulded into their roles by their acceptance of a challenge they read written in their history. It follows then that this leadership cannot survive the dissolution of the movement; once they leave the lighted arc of action, they descend into a historically unremarkable domesticity.

The presence of the people in Bankim's novels ought to be discussed carefully. Two points at least must be made. First, we must not erase their unmistakable presence; neither Rājsinha, nor Satyānanda, nor Devī, nor Sitārām would have become what they narratively were if they were not able to mobilize, through unspecified appeals, the great numbers who stand indistinctly behind and fight for them. In the novels, this crowd is not individuated, their faces are hazily, if ever, seen; but quite surely, they constitute an enormous 'absent' presence in these novels. Secondly, no doubt criticisms against Bankim's neglect of the popular could be advanced from anachronistic angles. But if we do not find a manifesto or manual for popular insurrection in Bankim, it is a fault of our expectations.

However, the endings of his novels of historical action constitute a more serious problem. In the question of the missing manifesto, we ask, why did Bankim not write like this? But here we are asking a question about narrative reality: why did Bankim write endings of this kind?

In the last four novels the contradiction between the climax and the ending becomes acute. We do not feel disappointed when, after a spell of individual heroism, Hemcandra leaves our narrative field of vision in search of an elusive small kingdom somewhere in the east. Pratāp's death, despite that great passage in which the voice of the author breaks in a tragic oracular performance, is still an individual sacrifice, and takes on no deeper philosophic meaning. For his death is not the death of an idea, or a movement, larger than himself. In the last novels, great imaginary actions rise to their climax only to come to strange but recurrent dismantling; Satyānanda leaves with the indistinct *mahāpuruṣ*; Śānti and Jīvānanda move away into a strange unspecified anonymity which could be death or domesticity with equal ease and irrelevance. Devī, more interestingly, returns to domestic life, but not without some signs of narrative struggle. She cannot escape her heroic identity so easily, and has to flee from it for the rest of her life, showing a certain objectivity of that image. Yet she is recognized from the back, the wrong side, by Sāgarbau, cleaning utensils at the domestic pond. Sitārām similarly moves away into obscurity, from the lighted stage of a few years. What is emphasized is the irrelevance of such non-political existence. They might all, like Devī or Praphulla, live a long and happy later life; but it is a different life, after a sharp rupture, and they are, like Praphulla, in a sense, fugitives from their own history.

Repeatedly, great imaginary actions come to this strange end. They are not defeated by their adversaries; they are defeated by their author.

The Double Endings

A literal reading of these ends would be false; and this would also go against the evidence of historical reception of these novels. The logic of the narration and the logic of the narrative may not be entirely symmetrical, or in a different way of saying, the logic of the writing and the logic of the reading; and it is possible for the writing itself to invite this displacement. This happens often in great art. And is it entirely accidental that long generations of Bengali readers have read Bankim wrongly, not accepting the endings in their literal form, but extending them in their reading? How can Devī, despite her narrative collapse into domesticity, remain a figure of resistance and hope? Are there indexes written into the texts which disobey or transcend the narrative conclusions?

Two points ought to be made here. First, for Bankim, there is a very serviceable doctrinal argument which could be taken from the *Gītā*, explicated in the famous phrase, *karmaṇyevādhikāraste mā phaleṣu kadācana.*[108] Actors have a right only to the performance of right actions, not to their consequences. To a modern sensibility this might offer an interesting doctrine of act—morality in a world of unpredictable, surely intractable consequences; a theory that moral judgement of acts becomes impossible and twisted if acts are to be judged by the consequences they lead to. Therefore, it is more cognitively justifiable, and charitable to the actor, to judge his act distinct from what it has brought about. However, Bankim is uninterested in this side of the implication of the phrase. What occupies his entire exegetic attention is a wholly different point, but vital in the context of the classical Hindu tradition of *naiṣkarmya.*[109] On a powerful interpretation of the tradition, the doctrine of *naiṣkarmya* advised abjuring action, at least of action which could generate attachment to its consequences. Of course, some traditions of Hinduism are quite tolerant to various mixtures of these injunctions, or relatively unorthodox practice of its tenets. Householders could, if religiously inclined, practise a life of non-attachment by not being seriously interested in domestic or commercial pursuits. Doubtless this must have given a religious aura to much simple domestic incompetence. For Bankim however it was this interpretation of classical Hinduism that had proved the undoing of his community. The meaning of moral existence must be displaced decisively. Instead of meaning a turning away from action, abdicating the undecidable moral quality of acts to God's infinite wisdom, it must mean, to this new view of Hinduism, an acceptance of responsibility for morally justifiable acts. Thus the person who would be known in the jargon of later years as the *karmayogī* (one who practises his yoga through and inside action) would ask for the right kind of action, acts which would be consequentially moral or justifiable. In a sense he would continue a principle of traditional morality in a reinterpreted form. Traditionally, naiṣkarmya was advised because *karma* or worldly action involved the principle of self-love. Selfish action could be abjured, and individuals could lead a non-world-directed life only if they withdrew from the world of normal action. In Bankim's interpretation of the principle, so crucial for Indian nationalist political models, the accent is shifted from other-worldliness, to being nonself-interested. If other-worldliness in itself is not the value which is pursued, but is an instance of nonself-directed living, this creates space for other possible interpretations. If there could be patterns of action which are not selfish, but in some defensible sense other-directed, then unselfishness need not always take an other-worldly form. Surprisingly, the nature of Bankimchandra's argument runs practically parallel to Max Weber's more renowned, but somewhat later version of a similar thesis.[110] Capitalism failed to emerge in India, Weber held, because although India had traditions of religious asceticism, these were primarily other-worldly, and counselled renouncing the actions of the world rather than their exemplary performance. In a strikingly parallel argument, Bankim asserts that Hindus, or in some cases Indians, failed to develop a sense of political nationalism and resistance against foreign invaders, because of their other-worldly asceticism. His positive doctrine, naturally, was to look for a principle of a this-worldly asceticism, particularly an asceticism of action. This moral principle was still an ascetic one, because it required an abdication of acts in the interest of the individual self. It was also ascetic because it advised frugality, exemplary economy of personal needs and pleasures. But the privation of needs was not to abjure the world, or because the world was unreal in some metaphysical sense. Rather, it was because of the fundamental significance of the social world, the significance and reality of its social and political order, and the imperative for its perfection. It was because the world was real, and its order must be improved and made less unjust, that people must undertake energetic but unselfish action. Unselfish here meant in the interest of a collectivity, of an extended self, a sense of a community which is still half-spelt, and half-seen in Bankim's times, which would later be called the nation. Thus there was a strong, ineradicable link between the coming of the nation and people acting in a particular way. The nation, nationalism, was not a mode of being, but of acting.

Therefore, the new doctrine of anuśilan, so wonderfully conveyed not by fallible actual figures but by the perfect actors of fictional discourse, is not merely a gospel of action, but also of giving up: men should be able to make sacrifices for action, but at the end, sacrifice the action itself. A doctrine of *karmatyāga* is announced at various climactic points of the stories—in the mahāpuruśa's instructions to Satyānanda after the battle, or the more droll abdication of action by Devī, or the tragic and complex betrayal in Sitārām. But I wish to argue that Bankim's texts are more complex than they seem. Some signals for a cancellating of this ending, a dream of these interrupted actions being begun elsewhere again, at other places and times come up at every end, so that these ends themselves take on an ambiguous and non-final character.

Figures of popular movements, though these are movements working in the fragile terrain of dreams, go back into domesticity, a trend that causes snorting indignation among Bankim's more radical critics. However, this appears to me to be an inattentive reading of the whole field of signs that Bankim creates and uses in his fiction, in fact, too literal a reading of the endings to be quite accurate. Domesticity is of course a field that is of extreme significance in Bankim's view of the world.[111]

Sānti is the active face of the feminine, as śakti, Kalyānī the peaceful, sustaining face indicated in an artless simplicity by her name itself. In a powerfully condensed sentence in *Ānandamath*, there is an intimation of a similar retreat into a fundamental domesticity, the everyday, unspectacular life of the people. This is a journey or a shift that Devī accomplishes within her own single biography: from Sānti she becomes Kalyānī. At the end of *Devī Caudhurānī* there is an explicit enunciation of the principle that domesticity is a sacred and solemn circle, a fortress, a place of refuge and strength, clearly a rejection of the ordinary babu dichotomy of a complex office and an insignificant home.[112]

> All this might appear extraordinary to others, but not to Praphulla. For she had practised the religion of non-attachment. After coming back to her home, she became a real renouncer. She had no desires of her own, she constantly looked for work to do. Desire means search for one's own happiness, work means search for the happiness of others. Praphulla was without desire, but constantly at work; that was why Praphulla was a real sannyāsinī (renouncer). Whatever she touched turned into gold.[113]

The last novels did enunciate the doctrine of anuśilan, of this golden touch.

Indeed, Devī's retreating inside this 'fortress' partly explains the invincibility of other women, all the way to the apparently comical victory of the critical wife over the handshaking rationalist.

All of these novels turn on a sort of narrative irony: the story ending twice, or in two ways, and the writer maintaining an ironical even-handedness towards them. Endings are both open and closed; they have one ending according to the authorial voice, and quite another according to the narrative voice. To put it yet another way, they have different logical and empirical ends. There is a tradition of literary and political interpretation of Bankim's art which sees in this simply political hesitation. But I suspect it is a better guess to see here an authorial device. Rājsinha's victory too is a victory of narrative. This is the weapon by which the historical story-teller gets the better of the actual process of events. By narrativity, the simple act of telling and not telling, the narrator can gerrymander events—not by directly falsifying them, but by the most unanswerable of his narrative weapons, his right of closure. In history, Rājsinha's victory against Mughal power must have been the merest of reprieves. In the narrative, Rājsinha's victory stands out indelibly. This difference is simply because history does not have an end, the narrative has. In history there were other acts which reduced the significance of Rājsinha's action; in the narrative, after Rājsinha's victory, there is nothing. By wiping out the beyond, erasing the historical sequel, the narrative can alter the significance of events without actually telling an untruth. It remains a truthful distortion of the truth of history.

If we look a little closer however there are all sorts of problems with the literal reading of the endings. In terms of intentions, a dismal ending of defeats simply does not square with the author's known intentions. These were novels of an uplifting message of praxis (anuśilan), and it would be ironical indeed if they ended showing the inevitable defeat of enterprises of defiance.

At any rate, it is evident that there is a paradox in the making and taking of the stories in these novels. Of course, the paradox can be given a simple resolution in one of two one-sided ways. It could be said, for instance, that the principle to be invoked is the one of authorial sovereignty. Seen thinly, the story does say that Devī did leave her role, Sātyananda did abandon his great project, and the text, it could be said, is entirely exhausted in saying this, its literal meaning. But this is too thin a theory. First, it takes away—illegitimately in my view—the right of the author not to mean what he says.[114] This illicitly assumes that the author has no right to set up a relation with his text other than of a simple, assertoric linearity. In the name of authorial sovereignty, this view takes it away.

Evidently, there could be a second simple solution, which would overstate the opposite case and say that Bankim wished to give the standard clarion call for nationalism. Actually, this reading runs into far greater problems, simply because to contest what is merely interpretative is easier than to contest what is literal. As an author he could have given simpler victory to the protagonists of anuśilan theory, a victory not in danger of being effaced. Surely, there was then something vitally wrong with his belief that all four were illustrations of anuśilan theory; after all, the literal line of narrative in at least three of them seems to indicate that anuśilan could not succeed without a favourable historical conjuncture. How does one account for, within a single explanation, the undeniable line of the story and its equally undeniable reception that appears to go against it so completely? Can one, or should one say that the reception of these novels is false or that it is true? An interpretation must also come to terms with the historical reception of a text which, as Jauss has argued so compellingly, reveals and constitutes its truth? Is the truth of a text to be constructed outside of its reception? Is a text so completely cleansed of its historicity a text any longer? Does not writing itself involve historicity? Is it not the ambition of every text to be moderately immortal and open itself out towards history and its attendant risks?

To understand the texts in their history, the apparent divergence of their saying and their meaning, and still restore a certain integrity to it, some kind of a distinction between two levels seems necessary. The first is the level of narrative syntax, how one event follows and relates to another, and what the adding of the last item does to reveal or displace the earlier configuration of the story. (This is most apparent in the thriller or detective story, but seems to happen in all narratives.)[115] *Faute de mieux* we could call the other level, one of narrative structure or ensemble—the coming together for configuration of crucial elements or signs, which override the narrative line, because these stay as pictures or gestures in the narrative unfolding. This is similar to a process that happens more discernibly in the viewing of acted narratives, in the theatre or in films. All created narrative objects require interpretation as constitutive parts of themselves. Stories in films or fiction are not objects that are distinguishable from interpretation, whole, uninterpreted objects to which interpretation comes afterwards. Interpretation is constitutive of them: to take an extreme position, a chapter following offers an interpretation of sorts of what earlier chapters have narrated; and there are some forms of stories in which the startling indeterminacy of the constantly recursive internal interpretations is what makes the distinct mark of the genre. Quite often, when the story is of a failed revolution, of a defeated protest, the poster would iconically thrust forward the radical gesture, a transient moment of triumph. In a sense that attempts to freeze the meaning of the story there while the narrative simply goes on.

We can thus make a distinction between the surface level of narrative where its literality lies, where the story makes its meaning in terms of its actual acts or states, and another deeper one. At the second level, the meaning is not made through a story in the same sense; actually, there is no story at this level, as structuralists would say. Instead, this is a level for the assemblage of act-elements or forms, and these are taken through processes of combination and recombination. It is from this level, apparently, that stories find the resources to annul what their surface says, to successfully carry a message which seems to be entirely opposed to the narrative, which yet passes clearly to audiences untutored in literary theory. Both success and failure of radical acts assert the possibility of radical acts, and create, to use the term in its Aristotelian sense, a radical rhetoric on the world.[116]

Let me now try to make this point textually. In fact, in Bankim's work, there are explicit markers in the texts which seem to invite this transposition, points where the text, by a shift of tone, shows the onset of narrative irony. I mean by narrative irony a deliberate position between two possible constructions of what the story says at its two levels, the literal and the figural. In ordinary conversation too, irony is often a matter of tone: 'x is y' means something quite different, sometimes entirely opposite when said in a different tone. The tone in this sense overrides the literal meaning of the sentence. There are similar devices in Bankim's narration. There are double endings not merely in the sense that endings can be read in two different ways: they are also in a more literal sense double. Stories end twice. And Bankim often uses the standard device of folk tales: dividing the narratively significant and insignificant parts of the lives of the characters by temporal or narrative disproportion. As in folk tales, heroes might, after their reported labours, live ever after in unreported (or unnarrated) happiness. Bankim's characters do the same. Rājsinha, we might hope, enjoyed conjugal bliss with his disproportionately young queen in such narrative oblivion. Devī spends what appears, by all indications, a long and productive life in her family after fading out of her fighting role. *Ānandamath* and *Sitārām* use a more severe style: nothing is said about Sitārām after the denouement; neither do we know anything about Śānti and Jīvānanda or Satyānanda; they are not allowed to detach themselves from their historical-fictional role.

However, this is not all. There are other interesting devices too. Consider *Devī Caudhurānī*, the novel in which these markers are the most explicit. Devī's story ends twice. After she decides to come away from her heroic role with Brajeśwar, when she abandons her role, that is the effective end of her tale. However her life story ends again in the last chapter. But the narrative irony is indicated in *Devī Caudhurānī* by a straightforward shift in the tone to story-telling. After Devī's decision, the narration suddenly changes from serious fictional prose to banter. When Devī comes to her village, someone spreads the word that Brajeśwar has come back with a fourth wife, a woman of advanced age. What follows is one of the great passages of farcical description in Bengali literature, describing in Bankim's inimitable language the utter confusion in the village caused by the curiosity about this new, aged bride.[117] Its combination of Bengali humorous idiom and observation of household foibles makes this passage untranslatable. Thus, the narrative irony is signalled by linguistic irony.

Alongside there is another device which is read everytime these novels are read, but which go unread in routine interpretations. After each of these endings, there are some curious sentences, quite unrelated to the narrative in terms of the novel form. *Rājsinha* ends on a comparison between William of Orange and the Mewar chieftain: 'In Europe William is renowned as the foremost among patriotic and righteous heroes; this country is bereft of history; so no one has heard of Rājsinha'.[118] In *Ānandamath*, after finishing the story of its protagonists Śānti and Jīvānanda, Bankim adds a characteristic line which confuses the tense and temporality of the narrative.

'Then the two of them got up, held hands and disappeared into eternity in that moonlit night. Alas, mother, would they come again?'

Devī Caudhurānī ends on a similar note:

Bhabani Pathak happily went into exile.

. . . .

Now come again, Praphulla. Come and stand in our midst—let us cast our eyes on you once more. Stand once more in front of this society and say, 'I am not new, I am old. I am merely those words. I have come so many times, but you had forgotten me; so I had to come again—

> Paritrānāya sādhunām vināśāya ca duśkrtām
> dharmasaṇsthāpanārthīya sambhavāmi yuge yuge'.[119]

These are proposals about time, the time in which the story happens and the time in which the story is read, and the common time, or history, in which these two times are held. The narratives end, but in each case there is an analogical extension: if there are similar circumstances in future, similar acts would be necessary. There is also a clear convergence of the two times: the future pointed at by the story is indeed the present of the reading. The tense of the report and the tense of reflection are different; but these powerful, passionate, unnoticed sentences bring them together into unmistakable convergence. The point of all this imaginary history is, then, the real one; the point of all this imagined past is the present. Meanwhile the world has remained largely what it was, made up of people and groups usually found in Bankim's reflections. But the great literary intervention into the picture of society had accomplished a task of great significance. Though the world and its inhabitants remain the same, its lines of distinction and opposition, its crucial lines of hatred and belonging begin to be drawn differently. So the political map of this social world has changed fatefully. Earlier, the Bengali middle-class babu considered it self-evident that the basic line of demarcation ran between those who had reason and those who did not, an arrangement which classified him, together with the English, on the same lighted side of humanity. This great critique of his pretensions, the insistent, remorseless destruction of his discourse of self-deception shows that the crucial line is drawn at a different point in the social world, between rulers and subjects, and thus, it pushes the intellectual to the surprising discovery, which he mistrusts and resists, that his natural fellow beings are the ordinary people. By shifting the line of distinction the author has pushed the Bengali babu of the nineteenth century from the community of reason into a community of discontent.

After all, the writer of these subtle endings, of this indefinable time, is Kamalākānta Cakravartī, in his altered and transcended self. It is not easy to contain or defeat this master of the dissembling discourse, of saying the unsayable. As long as it is the terrain of discourse he is sovereign. It is not surprising that he could create just such novels with just such endings, such poetics which make it possible for the narration to end but the narrative to go on beyond its end, that it stretches itself in such deep and majestic anguish from imaginary history towards the real one.

5

Tragedy, Irony and Modernity

> Irony could not develop before the age of reflection, because it is constructed from something false which by means of reflection disguises itself behind a mask of truth.
>
> Vico

A civilization develops over a long history something like a grammar of vision, an optics on which its aesthetic practice is based. Artists and their audiences see objects in the world, the world itself, its various possibilities through that grammar of seeing. This can perhaps be clarified by an analogy with painting. It is like a palette of primary colours which constitute a structure which restricts and forms possibilities at the same time. Traditions of high culture develop in this manner a palette of feelings, sentiments, ways of describing, feeling, perceiving things of the world.

There is no universally admissible, neutral manner of seeing the social world, the view from nowhere; there cannot be a view that is not of a certain kind. This optics of the literary field is created by aesthetic traditions—usually a complex mixture of explicit and unannounced preferences, ways of shaping things by giving forms, names, destinies which are always at work within the work of authors. It goes without saying that though these are deep structures, with all that this implies, these are not beyond history. Historical requirements affect and transform them, though these demands are neither constant nor linear, happening everyday or in a specific direction. Aesthetics are in the nature of structures of resources which are always available for individual crafting; and through the pressures of such crafting of resources to demands, new items are added to aesthetic structures or their forms are changed. The celebrated Indian term for this grammar is of course the theory of rasa: Indian aesthetic uses a palette of the nine primary rasas as its basic generative structure. But the grammar is a historical thing and presupposes a world. The theory of rasas emerged from reflection and perception in a certain kind of world, where objects and possibilities were structured in a specific [illegible] the world that emerged from colonial contact, and the sensibility that this generated was fundamentally different. It is this peculiar position in a world that is being historically reconstructed which forced reflexivity upon artists and writers of the period. Bankimchandra was an artist of tradition in a very technical sense. His palette was a traditional one of the primary rasa structure. Often it is easy to describe his tropes, the specific literary operations taking place within his texts, entirely in terms of traditional *alaṇkara* terms—*upamā, utprekṣa, yamaka, vyajastuti,* even the most conventional of these stylistic features, the *anuprāsa.* Yet his art was entirely different from that of his successful contemporaries many of whom had chosen objects, styles or compositions which, in a world of invading modernity, still sought to retain a whole, unfractured aesthetic tradition. Turning their backs to the historical changing world, they could take refuge in traditional composition. Some, like Vidyasagar, experimented with a new language, though still seeking legitimacy from its obvious kinship with sonorous Sanskrit. But none took the risk, the enormous danger of turning round, and facing the world of modernity with that traditional palette in hand. Vidyasagar's objects of aesthetic depiction and narrative were Sītā, Śakuntalā, figures who were traditionally depicted by that palette.[1] Thus the characteristic poses, the moods, the emotions suited the palette. Bankim chose figures of a different kind, either living in modernity or thinly living in the past but already experiencing modern forms of suffering, a whole world of emotions, moods, poses, situations that were not formerly the task of this palette to analyse or portray. I think that was why Bankimchandra's art could find no serious imitator or successor,[2] but still has the power of all genuine origins, their ambiguity, their mixture, their duality, their intrinsically transient character. He was neither of this world, nor of the other; yet his art was in a sense adequate to the complexities of both. His aesthetic can be set against that of classical Sanskrit literature, and at the same time, that of the modern. For most other artists in the Bengali language such a double comparison would be wholly inappropriate. Vidyasagar's characters experience traditional emotions of *vātsalya, karuṇā,* etc., not the indescribable emotions of modernity. Tagore's figures not only experience the new emotions, but have found a new language in which to adequately express themselves. They have been given a new palette. Only in Bankim the world is new, the palette is old. And this transient adequacy gave to his art an indescribable and incomparable dignity. This is why Bankimchandra fashioned the Bengali sense of modernity, because he was the first to teach them to wonder at this phenomenon of history reflexively. He was the first to create something like the rasa of modernity, for which the earlier primary colours of the palette of feelings was inadequate. To contend with the structure of feeling that modernity created in this world slowly growing aesthetically unfamiliar, he had to mix and experiment with rasas. The eventual rasa of modernity, if we can so call it, is fundamentally tragic, and it mixes two primary colours which traditional aesthetics never brought into contact. These are the moods of tragedy and laughter.

In terms of narrative form as well, Bankim relied quite openly and heavily on the resources of traditional Indian aesthetics. His narratives were called novels, but were actually structured more like traditional dramas or narratives. His irony of the babu and the world of colonial powerlessness uses constantly and with consummate skill the earlier forms of the vyajastuti. Even his material depictions of nature, or the physical description of women, come straight out of an aesthetic canon which goes back into the ancient texts and is at the same time solidly present in the folk optics of nature and femininity, both *prakṛti* to classical theory. At the same time, the taste of Bankim's aesthetic is unquestionably modern. In Bankim's efforts to create this new aesthetic the most significant element, because it is also the most contradictory, is his use of irony. Classical Sanskrit and early Bengali literature[3] had long and powerful precedents of poetic compositions of intricate irony. He assumed and extended this line in one sense. Yet, what he did with it, the way he extended and added on to it, the way he made it perform new functions of meaning are astonishingly modern. Irony and self-irony are indeed very close and enormously distant. The literary or stylistic techniques are common, the rasa is radically distinct. Self-irony is both a crowning achievement of irony and an overturning of its conventional form.

Traditional societies often find the identity of others puzzling, to puzzle about the self is an unmistakably modern form of reflection. Settled traditional societies were marked by an assumed sense of self which was rarely a matter of debate, and except in some rare occurrences, never a matter of personal intellectual search or deliberate construction. To construct an identity is an essentially modern activity, and it is part of a history which offers various possibilities of existence for individuals and groups. In a fundamental sense both individuals and groups can, under philosophical conditions of modernity, choose what they would like to be.

The modern absorption with the self has some interesting and contradictory aspects. Thinking about the self is always mildly narcissistic, it gives consideration of the problems of the self great significance; but it also implies the necessity of a certain detachment. To look at a face in a mirror is still a project of distancing and detachment; and detachment implies in its train the existence of different possible ways of constructing the self and a life that tries actively to fashion the self's form.

Formerly, intellectuals often described and reflected about why other people were different. European travellers and chroniclers did not merely observe the strange customs of other cultures, but also made interesting hypothesis about its reasons. Al Biruni, to take an outstanding example from the Indian tradition, speculated about the Brahminical ways of the Hindus. But typically, neither the Europeans nor the Arabs or Indians worried about what they were themselves or presented it to themselves as an intellectual problem. Being something was taken for granted, at least not considered problematic; it was the being of the other, how people could be different from what people should normally be like, which raised problems.

To turn the self into a problem, either philosophically or sociologically, is a wholly modern idea. And to discover the self is a potentially tragic process. Through the process of reflexiveness, the self is tormented by desire and failure, by its increasing sense of inadequacy, assailed by desires to become something other than what it at this moment is. It can be claimed that Bankim is the first writer in Bengali, and surely one of the first in any Indian language to turn the self into a problem and investigate its possibilities of construction. He is also one of the very early thinkers to appreciate its historicity, the fact that people can become what they wish to be, and not merely stay what they immediately are.

I wish to suggest that it is this reflection on the self as a construct which directs Bankim to his distinctive style of humour. It is a commonplace that Bankim is one of the great writers of humour in the Bengali language. But humour can exist in literary works in many different forms; and to understand what it really does for a particular author it is important to find out in exactly what specific way it exists in his art. In both the literary traditions that he knew, variety is greatly applauded. Modern western fictional traditions which Bankim knew encouraged the mixing of humour both in the seeing of the world and the telling of the story. Classical Indian literary doctrine was, if anything, even more explicit about the great merits of mixed compositions. The theory of the *Nātyaśāstra* for instance encourages mixtures between different primary artistic modes like dance, vocal music, drama. It also encourages the mixture of the rasas with the injunction to respect the difference between the *sthāyī* and *vyābhicharī*, without which the work of art loses coherence of signification. Yet those mixtures were meant for pleasing the audience, enhancing the attraction and aesthetic value of the object of art. Bankim's humour arises from a different, and in my view, much deeper source, as the only recourse in his personal and historical predicament. Traditional theory uses fine distinctions between different aspects or forms of *hāsya—upahāsa, atihāsa.*[4] Implicitly, this sort of differentiation of the rasa deploys an implicit norm of moderation, because *atihasita* is a mode of excess, whereas *upahasita* is a subtler way of using the margin of signification that is rendered possible by restraint, by way of understatements. Still, despite this enormously impressive palette of the hāsyarasa traditional forms of wit and humour were insufficient for Bankim, for the rasa or bhāva of his historical and biographic experience, of loneliness, incomprehension by his contemporaries, the peculiar melancholy of living in an alienating and intractable history, of an overwhelming feeling of unfulfilment, and the sorrows of discovering the self. It is this predicament—of a self which could not be justified but yet could not be condemned simply because it is too close, inalienable, too dear, simply because it is the self—which makes him look for techniques of transforming the earlier aesthetic of humour into the new, somewhat darker form. Bankim in many ways invents self-irony as the mode which is aesthetically adequate to modernity. Although Bankim made some crucial moves towards the formation of nationalist discourse, its object was still tentative, provisional, fragile in his imagination. He could not draw from it the kind of assurance that later nationalists drew, which meant that they did not require irony to the same extent. They had sorted out the contradictions, or could robustly pretend that these had disappeared. Thus given their disposition to the social and historical world, they did not require irony towards the self; indeed, they suppressed any irruption of doubts as to the seriousness and justifiability of this enterprise.

It is a commonplace that Bankimchandra was one of the creators of the imaginative world of the modern Bengali. He does not look very successful if we try to count his followers. He led no movement. He founded no church. He did not announce a new political utopia. His contributions to the forming of the Bengali world of modernity is, I have tried to argue, far deeper. Though he did not shape ideas themselves, he shaped the medium with which ideas are made. His own perception of modernity might not have found favour with many; but he gave to the modern Bengali the palette with which it became possible, for the first time, to paint intellectual and imaginative pictures of modernity. His contribution to Bengali literature is well known; what I have been concerned to argue in this work is that he inaugurated a modern, mature form of historical self-reflection, and he shaped a language which had the richness, subtlety, variety, suppleness to engage in such a task; and since then, though transformed several times by others, this language has remained the most fundamental, if also the most invisible, instrument through which the Bengali has tried to cope with the mysteries of living in the peculiar historicity of modern times.

Inevitably, because of their historical context, most thinking Bengalis of his generation reflected on the causes of colonial subjection and the nature of British rule. Only a few, like Bhudev Mukhopadhyay, had the breadth of vision and the philosophical temperament to turn this towards a theoretical reflection on modernity itself. Between the two of them, who were contemporaries, Bankimchandra's reflection had a greater dramatic quality, because of his gradual revision of his earlier enthusiasm for rationalist doctrines: he had a most agitated, tense, self-correcting self to contend with. Bhudev was saved from the necessity of irony about the self by the relative stability and consistency of his views. Since Bankim changed his own views, there is a subtle necessity of tempering the criticism by humour, because on both sides of the ideological divide, in the person who criticized and the person who was the object of this criticism, between the man who attacked and undermined and the man who felt undermined, in a sense, was the same self, in its two temporal versions. His youthful views were not rejected by another person, but by the necessarily ironical mode of a later self. His irony is the direct opposite of Vico's: it could not have arisen before the age of reflection; but inside it is truth which expresses itself by assuming the disguise of falsehood.

Although irony is so evident and remarkable in Bankim's art, its predominant theme is not an engagement with the small embarrassments of the everyday but the movement of history; and his general sense of history is essentially tragic. Evidently, this kind of irony, though a form of humour and depending on the formal aesthetic of laughter, is quite compatible to the tragic sense of history so noticeable in his reflections on modernity. About modernity's dimensions, structures, and trends Bankimchandra's historical thought is admirably clear and unhesitating; but interestingly, it does not pretend that it is at all clear about the future. Bankim's considerable reading of contemporary European social theory—he was very familiar with Comte, Mill and Spencer, and knew of Hegel and the French philosophers—had not persuaded him to accept any of the great philosophies of history and the assurance these uplifting narratives provided in face of the fragility of things that modernity had brought about. His acuity of analysis of the present did not tempt him to declare that he saw the future with any degree of clarity—something that later nationalists of much less historical perspicuity announced without hesitation. Bankimchandra was therefore left with a peculiar vision of history. History was of the greatest significance, because all things social were malleable, in flux, rendered provisional. It was consequently necessary to achieve a rational understanding of the process of the present as history. Yet it did not yield any clear picture of what would emerge in future, except images of improbable utopian colour, captured in the complex symbolism of his obsessive dreams. There is particularly one sequence which appears often in varying forms, as a dream of some of his fictive characters, as a dream of his secret self—Kamalākānta. This is the image of a splendid golden icon tossed in the dark waters of time, again a traditional image, but most appropriate to his sense of modernity. History, the way he reflected on it and made an object of serious reflection, was a wholly modern theme; but he had successfully derived from the great stock of images of the classical past its most appropriate symbolic image.

On the darkness of history

In Bankim's writing, the question of British colonial administration was broadened, at the best points, into a question of modernity in general. And the question of modernity broadened, at the best points of his performance, into a philosophic reflection on time. It has been argued persuasively that modernity alters the nature of temporality.[5] From a cyclical repetitive ontology of time, European theoretical thought and later popular consciousness emerges by degrees into a conception of future time as an opportunity, as a broad field of creative activity in which natural and social things can be shaped according to collective human desires. The problem was not so much the plasticity of the social world, but rather the difficulty in arriving at some commonly accepted decision/project about what to do in that time.

In a very general way, modernity brings in an identical change in India, from a time that is given to a time that is made, i.e., formerly, time was not seen as an opportunity in which well-conceived social or political projects could be carried through. Individual lives have to be lived, in dignity and humility, within terms which are forever set, over which human beings have no control, not to speak of sovereignty. Over a period of about a hundred years, this wholly passive attitude towards time changes into a conception far more active, occasionally superficial, oversimplified, hasty, unintelligently optimistic. But the steps and exact context of this change were very different from the narrative of modern temporality in the West. The coming of the new time could not appear the same to Indians for two reasons. The optics or ontology of the world through which they saw the coming of modernity was different in conceptual terms. Besides, the content of this modernity, what it unfolded in front of their interpreting minds, was also quite different. While modernity in Europe was primarily an experience of liberation, in India its main element was a narrative of subjection. Bankimchandra's work is interesting precisely because he shows how the earlier optics of time slowly gives way to the possibility of another.

In traditional Indian thought, time was a dark god, sometimes insufficiently distinct from death. Indeed, there is a peculiar indeterminacy, ambiguity in Indian thinking about the precise kind of reverence that is to be accorded to time. Time is evidently irresistible, its ravages inescapable for human lives; but this elicits a response that is more of awe than of regard. Time, *kāla*, is thus given a peculiarly fearful reverence, because of its affinity to death. In traditional Indian consciousness, the metaphors of time are chosen in accord with this predominant sentiment. Time is a river, but a dark river, carrying everything towards an inescapable destruction (*laya*), finishing annulment, non-being. But this river of time carrying everything along does not surprise people with destruction. It does not do anything sudden, episodic, frightful in that sense of a catastrophe: it is the inevitable definition of life as a being towards death. The waters in the river of time are dark, a metaphor appropriate in two ways: the traditional meaning was that time represents intractability, impenetrability. What it does is beyond control and comprehension. In Bankim, history is often related to this river of time. The dream sequence in *Kamalākānta* reveals what lies as it were in the cultural unconscious of the Bengali intellectual: the two most intense modern experiences of history and colonial subjection appear in this dream through their most appropriate traditional metaphors, of the mother goddess and of the dark waters of time. Kamalākānta sees, under the influence of opium, when truth reveals itself to him in the illusion of dream, that the idol of the motherland, golden and immeasurably precious, is being swept away in the terrifying dark river of time. Some new elements can be discerned in this use of metaphors—displacements to deal with the peculiar features of modernity. Formerly the river of time was merely dark, representing a cognitive impenetrability, and a helplessness that all creatures share in. Now there is a clear suggestion of the waters of time getting turgid and muddy, so that the figure of the motherland is subjected to a greater defilement. And its helplessness is not the frailty of all creatures in face of time, the intimation of mortality. It is a more cruel desertion by destiny, a fate of subjection to the power of others. Death is a great calamity, but it falls to everybody and is therefore a great leveller. Subjection, by contrast, is a bitter fate that falls to some, not to others. It causes a strange kind of bereavement, a strange and partial death; a people can observe the bereavement of its own liberty, mourn, paradoxically for itself.

In another way too Bankim is a figure hesitating at the gates of modernity, of traditional and modern story-telling and fictional aesthetics. Indeed it is possible to argue that there are two fictional aesthetics struggling in his art, of very different kinds, with very different ideas about what depiction in art is about, and who are the figures and actions that deserve the celebration of art. In most traditional cultures, such a figure would be heroic: of a type who is entirely in harmony with the values and ideals of his society, and willing to make great sacrifices for them. It is this form of extraordinariness of individuals and their lives which were celebrated in traditional art. No doubt Bankimchandra himself created a number of such figures in his fictional art, though the world in which they live is almost always marked by some complexity. They are not in consonance with a world already given, but a world to be made and whose sanction already lies in the desire of people around them. Of course this too involves an idea inextricably connected with modernity, of the plasticity of the relations out of which the social world is made. These heroes create a new national order, rather than reestablish one which had been temporarily ruptured; though, without doubt, this is often accompanied by a restorative rhetoric.

But I find something else in Bankimchandra's art which is even more decisively a product of modern consciousness: the perverse attractiveness of Kamalākānta as an extraordinary individual. *Kamalākānta*, I have argued, is quite central to Bankim's art, because it puts the question of colonial powerlessness at the centre of vision, as a nagging, unforgettable fact. But his figure is interesting not only because of his subtle and subliminal politics, but also for a purely artistic reason. If artistic attention is the mark of a hero, then Kamalākānta is certainly one in Bankim's fiction; and it is also true that the whole business of the world in the *daptar* revolves around him. The world unfolds to us through his eyes. And he is not only the master in this world of words, in the fictional world of action, in the court drama too he is the central figure, because he faces the court in a doubly unequal battle. On one side, he is a single, entirely lonely individual against the whole structure of colonial justice. On the other hand, he is not alone, he has the support of a huge historic civilization, all its subtlety, intelligence, evasiveness, refinement is given to him to exploit and embody for this single occasion, because he has made the terrible choice, to be its representative, against all odds. On this level of course, his adversaries are unevenly matched with him: their culture is shallow, imitative, insecure, inauthentic. He answers their questions with the assurance of an intelligence of a thousand years.

The astonishingly modern quality of Kamalākānta lies in the fact of his being an anti-hero, a hero who is a congeries of totally negative qualities, but this heroism being still an act of deliberate choice. In fact, he accomplishes the most amazing series of displacements in the aesthetic of fictional figures. Kamalākānta is not a success in any conceivable sense at all in terms of earlier, more straightforward aesthetics. Indeed, he is unlike the heroes typical of traditional narrative forms; unlike the colourful figures of Bankimchandra's own fictional works; unlike the more problematized heroes of later high Bengali novels including Tagore, unlike again some of the types created by Saratchandra Chattopadhyay drawing upon the slightly inverse heroism of the terrorists—at once mysterious, dangerous, attractive to the middle class. The obliqueness with which Kamalākānta stands is much more radical, unamenable to the standard technique of being redeemed by the love of women, representing the disinterested, uncorrupted higher intelligence of this society, as happened to the terrorists in the novels. Though shunned by middle-class society, their ability to attract the admiration of women showed the possibility of standards by which they could be redeemed to more straightforward norms of fictive glory. Their return to unremarkable domesticity, the site of normalcy and acquiescence in norms, is simply narratively unreported. Kamalākānta is unamenable to this mildly deviant formula of Bengali fiction; his obliqueness, his ability to remain unreconciled, is far more radical than the practical violence of the terrorist figures, than their cult of radicalism of the act. Kamalākānta is wedded entirely to the principle of radicalism of thought, the radicalism of daring to think unlike everyone else. He is the single great hero of uncompromising critical thought that high Bengali literature of the early period produced—desolate, incomparable, inimitable in his peculiarity, ceaselessly lamenting the fate of subalternity, reflecting the darkness of history, but not allowing that to reduce his enjoyment of the comic spectacle of the everyday. He lives in a world of laughter framed in a subtle and unstated sorrowfulness. He is a character in Bengali literature without a forerunner or a successor, and in that sense the most interesting figure in Bankimchandra's art. To understand Bankimchandra there is no other way except to understand the laughter and darkness in the world of Kamalākānta.

Can we make some coherent sense of all that I have said about Bankimchandra Chattopadhyay? What kind of a person was he? What sort of historical discourse did he live inside? Historical remembrance is a notoriously mixed fact: being remembered is a process of the accretion of meanings on to texts. Texts are not merely set in history in a broad sense, they also set in motion little histories themselves. What sort of history did he set in motion, or contribute to? It is often seen as an obligatory task of the history of ideas to look with persistence for the 'coherence' of a mind. It is interesting to see why minds like Bankim's, with their opulent gifts, found it hard to decide to be coherent.

All history is contradictory, but not in the same sense or to the same extent. The history of colonial societies is ridden by contradiction in particularly fundamental ways. The puzzlements of colonial intellectuals arose from the unmerciful way history offered them the gift of modernity, in a way in which it was inextricably related to a destiny of subjection. Unlike as in Europe, modernity came to India as a primarily external proposal as a theory and an external agenda as practice. The Enlightenment, for all its later complexities, had begun with a 'happy' history in Europe. Its authors were happy writers, for its protagonists saw the processes of modernity as one which spread liberty across social life. The historical situation of the colonial writer was tragic because of the unjustness of the choices facing him. If he chose modernity, he had to choose subjection as its condition, or so it appeared to him. If he chose autonomy, he had modernity as a necessary price. These two positions developed into two separate discourses, which made exchanges across their boundaries increasingly difficult.

Most intellectuals decided to think in and through one of these discursive forms: the Young Bengal with modernity, indigenists with the discourses of tradition, and in the peculiar cultural history of colonial Bengal, it required a certain kind courage to speak either discourse. But for both there was at least the assurance of an appointed discursive space, with their common arguments, languages, and set audiences. Some, like Bankim, failed or refused to make the choice, and saw the structure of this choice itself as tragic. Their intellectual positions were bound to be more lonely than the others'. Hemmed in between two large circles of discourse which sought to deny them any space, what they said was constantly in danger of being assimilated into one of the two dominant forms. They had to struggle constantly against such assimilation. Their inability to turn decisively away from the two sides of their history, their continual interrogation of both, condemned them to a form of 'unhappy consciousness'. Their solution was to remain contradictory in their consciousness, for to be consistent was not to be open to all the human possibilities that history had placed before them.

The gift of history, bestowed by a modern western education, appeared to them a deeply ambiguous one. They had learnt from it to think of their lives—singly and collectively—in the frame of linear time, and in terms of the rational capacity to construct the present and future. But in their condition, this made a self-conscious existence more difficult. History must have appeared to them a strange mixture of disclosure and concealment captured in Hegel's phrase, the hieroglyph of reason.

There is a peculiar relation of adequacy between the experience of this history and the kind of literary art that Bankim created. Alternating between a tragic fiction and comic commentary, a view of the world filled with darkness and a view of the everyday filled with laughter, attracted to both rationalism and its critiques, both enlightenment and religion, tormented by the choice between subjection and modernity, he had discovered a way of dealing with all this complexity in irony. Irony allowed the contradictions of the world to be retained without recourse to a reducing consistency. Between its solemnity and travesty, its darkness and laughter, he had fashioned a language that was adequate to express the sense of the world as it appeared to his particular unhappy consciousness.

Notes and References

Chapter 1

1 Martin Heidegger, 'Origin of the work of art' *Basic Writings*, Routledge and Kegan Paul, London, 1978, 170.

2 I do not deny that the world within the text depends in very significant ways on the world outside; but in literary studies the two things must be seen distinctly. An excessive emphasis on the idea that the artist is a voice of his times could make it impossible to see the specificity of the artwork.

3 There is considerable literature on this problem; a recent and interesting enquiry into these questions in Arthur Danto, *The Transfiguration of the Commonplace*, Harvard University Press, Cambridge, Mass., 1981.

4 Cf. Danto.

5 I have tried to interpret his humour in the next chapter.

6 For one definition of liminality: 'During the intervening liminal period, the state of the ritual subject becomes ambiguous, neither here nor there, betwixt and between all fixed points of classification . . . in liminality, the symbolism almost everywhere indicates that the initiand, novice or neophyte is structurally if not physically invisible in terms of his culture's standard definitions and classifications'. Victor Turner, *Dramas, Fields, and Metaphors*, Cornell University Press, Ithaca, 1974, 232. The sense in which I use the term differs from this use. The common point is that I also see the liminal as non-standard. However, I see the liminal as being outside, and not as a point that is transitory between two standard states.

7 Novels which raise such questions quite clearly are *Durgeśnandinī*, *Kapālkundalā*, *Mṛṇālinī*, *Viṣavṛkṣa*, *Kṛṣṇakānter Will*, *Jugalāngurīya*, *Rādhārānī*, *Candraśekhar*, *Ānandamath*, *Devī Caudhurānī*, *Sītārām*—11 out of 14 of his novels.

8 Jagatsinha, a young commander in the Mughal army, is developing a relationship with Tilottamā when he is captured and injured. When he regains consciousness, he admits that he had confused between Tilottamā and Āyeṣā, daughter of an Afghan chief. This symbolically emphasises the indistinctness of the two women.

9 Although Bankim mentions Tilottamā as 'Durgeśnandinī Tilottamā', *BR*, i, 9.

10 Tsvetan Todorov, *The Poetics of Prose* (trans. Richard Howard), Basil Blackwell, Oxford, 1977.

11 The conflict arising out of a human attraction across a social line of prohibition is clearly emphasized in *Durgeśnandinī*, *BR*, i, 12–13, 15.

12 Passion has its own justification. Given this, we come to confound things—good appears like evil, evil like good; good and evil appear indistinguishable and difficult to identify. Several explicit 'inner dialogues' occur in Bankim's fiction; e.g. *Kṛṣṇakānter Will*, *BR*, i, 499.

13 The classic arguments about the problematic hero appear in Lukacs, *Theory of the Novel*, Merlin Press, London, 1978, and were pursued by others like Rene Girard, *Deceit, Desire and the Novel*, Johns Hopkins University Press, 1976.

14 *Mṛṇālinī*, first published 1896.

15 Hemcandra is very close, in terms of character type, to Jagatsinha, the insensitive hero of the first novel. In his early novels male figures fail to achieve even much villainous complexity. This is not true of the genuine tension in the figure of

Paśupati, whose fortunes and initiatives hold the story together, while Hemcandra does most of the manual work.

16 *Mṛṇālinī*, *BR*, i, 166–7, and the ninth chapter in general, 'Mohitā'.

17 Paśupati's theory of politics assumed that rationally calculated moves attain their objectives. This did not take into account several things: (i) that others may pursue motives that one may fail to properly read, (ii) that there were unintended consequences of rational actions, and (iii) that in politics large popular forces often decide what happens, upsetting immaculately laid out rational plans.

18 Astrological forecasts have always two layers. At the first level, there are forecasts made in a very general form, at the second level there is the task of reading this generality into individual incidents in individual lives. The story of Mṛṇālinī is particularly affected by these falsely true horoscopic predictions. Hemcandra's guru predicts—correctly—that Muslim rule in eastern India will end through the initiatives of 'merchants who live in the west'. This is falsely read by the same guru as indicating the imminent success of his disciple who had sometimes disguised himself as a 'western merchant'. Another prediction says that Manoramā would die, when very young, on the funeral pyre of her husband, which is what comes to pass, despite her father's attempt to ward it off by concealing her identity. Similar forecasts occur in *Candraśekhar* , *Sītārām*, *Rājsinha* etc.

19 I realize that this is generally, not literally, true. A woman cannot be a wife and mistress at the same time (to the same man). Purely logically, she can be mistress and wife sequentially—if she is first mistress and then wife; the reverse of this seems impossible for Hindu order, because divorce is prohibited. In a curious and odd way, however, Indirā's case is one of first wife, then mistress—which precisely is the point of the story.

20 Sigmund Freud, *Leonardo da Vinci*, Random House, New York, 1966.

21 Significant philosophical consequences follow from telling a story by four selves, instead of one narrative *self* and three *others*. Bankim however is generally uninterested in such questions, and seems to have tried this form mainly for the sake of novelty.

22 In Bengali there was a clear distinction between what were usually known as *sādhu* and *calita bhāṣa*, differentiated especially clearly by the differences in verb forms. Any mixing between these high and low forms was denounced as *gurucaṇḍāl doṣ*: symbolic mixture of the highest and the lowest castes.

23 By this I indicate what seems to me a simple but useful distinction between the language of the narrator and the language used by those who figure as characters in the narrative.

24 Bankim is openly contemptuous about the low colloquial, what he calls the *hutomī bhāṣā*, *BR*, ii, 200.

25 I am setting out elsewhere an argument that Bankim regarded language as a theatre of power.

26 It would be a big and crucial mistake to believe that the colloquial Bengali of the kitchen is a mere repetition of the bhadralok babus' of the *hutomī* kind. Bankim's contempt for the *hutomī* language is entirely justified because it is based on an implicit distinction between two quite different colloquialisms. It is also linked to a social or cultural distinction between the male and the female world for at this point of historical evolution, the woman in the bhadralok household has not become a babu. The colloquial language of the babus' *baithakkhānā* is a vehicle of idle, often prurient gossip; the colloquialism of the kitchen is not obscene, it is close to the earthy practicality of the popular. Obscenity is not a product of the popular imagination and lack of refinement; rather, it is a product of the degeneracy of the literate.

27 See the hilarious and entirely predictable exchange between a babu back from the office at the violet hour and his wife, whom he finds reading a Bengali novel, which turns out indeed to be Bankim's own *Viṣavṛkṣa*, 'Bāṅgālā Bhāṣār Ādar', *BR*, ii.

28 *Kamalākānter Jobānbandī*, *BR*, ii, 104.

29 Examples could be found in the conversation between the babu husband and his unbabufied wife in 'New Year's Day', *BR*, ii, 47, in which the wife destroys the solemnity of English etiquette by a devastating colloquial redescription in the vernacular.

30 To attribute Sītā's rejection by Rāma in the epic simply to *Lokāpabāda* or anachronistically in modern terms, to his male hardheartedness appears too simple. There is a problem of the impossibility of proving to the hard (though perhaps malicious) sceptic that Sītā is innocent, and above reproach.

31 Which is similar to this sort of second order questions about morality I find in *Durgeśnandinī*.

32 Its oblique comment on Hindu sexual morality is also interesting. A husband takes his innocent wife back into the marriage relation. Morally this is commendable. But he does this not out of any virtuous intent, but due to his inability to live without her—the worst of all possible sins in Hindu morals. It is irrelevant that the consequences fall so beautifully for everyone, especially for him. By his actions, he actually deserved the worst.

33 In the Hindu tradition too sometimes desperate expedients are sanctioned: *vivāhakāle ratisamprayoge praṇātyaye sarvadhanāpahāre* in one famous sloka.

34 *Mṛṇālinī*, *BR*, i, 'Antimkāle', 204–6.

35 Though there may be a connection here between the picture of decline and devastation in *Mrnālinī* and the regenerative constructions of the later, more mature historical novels, in a stylized form one could say that here is the *problem* symbolically presented and those are a symbolically stated *answer*. See below, chapter IV.

36 In several other novels however—perhaps inadvertently—Bankim plays on this theme of superimposition of identities. In *Candraśekhar* Śaibalinī is placed in liminal situations—she becomes an outsider to her social role. She is of course in a dramatic and fictional sense more complex than Indirā. At some points however she shares in a general waywith Indirā or Lutfunnisā (*Kapālkundalā*) the attribute of going outside of standard definitions.

37 Exactly like the prophecy about Hemcandra, this too is a two-level prophecy—the first couched in general terms which leaves open the difficult second level of its application to an individual destiny. Śrī is shown in her horoscope as *priyaprāṇahantrī*—one who would cause death to someone dear to her. Sītārām's family conclude that as the dearest person (strictly, the most beloved—*priya*) for a woman must be her husband, Śrī was destined to cause her husband's death. At the end of the novel it turns out that she causes death to her brother.

38 Cf. *Sītārām*, Chapter 9, *BR*, i, 880. Rāmcānd and Śyāmcānd are, I suspect, babus anachronistically interloping into an earlier century. There are many such anachronistic characters in Bankim's novels—babus misplaced, for partly humorous, partly didactic reasons, in other times.

39 For recent analysis of this point with references to *Ānandamath*, a novel I have generally neglected in this chapter, (because it is excessively well known, imparting a misleading bias to the popular image of Bankim's literary art), see Jasodhara Bagchi, 'Positivism and Nationalism', Review of Women's Studies, *Economic and Political Weekly*, 26 October 1985.

40 'Depart, then, Pratāp, for your residence in eternity. Leave for the place where conquest of the senses does not cause suffering, beauty does not cause desire, love does not lead to sin. To a place where there is infinite beauty, infinite love, infinite happiness, and an infinite virtue in being happy. Where each one understands the sufferings of the other, each preserves the virtue of others, obligations of others, where people sing others' praise, one does not have to die for another's sake; leave for that place of infinite grandeur. Even though you get a thousand Śaibalinīs at your feet, you would not have the desire to love them'. *Candraśekhar*, *BR*, i, 424.

41 In the story Pratāp and Śaibalinī deserve each other, for both are handsome, are in love and grew up together as children: their love is natural. Social conventions rule out their marriage, and they are married against their own inclinations, against their own natures, to Sundarī and Candraśekhar.

42 We must remember that Bankim occupies a position midway between two philosophic traditions, between eastern and western philosophic theories. In his work, reason has a shifting, sometimes double, definition. Rationality in the Hindu tradition, unlike in western philosophy, represents a certain stillness or fixity, and end to agitation; a classical example of this would be the definition of the *Sthitaprajna* in the *Gītā*.

43 In Bankim's fictional universe the marriages of Nabakumār, Matibibi, Paśupati, Indirā, Nagendra, Govindalāl, Candraśekhar, Mubārak, Mahendra, Brajeśwar and Sītārām are broken some way or other, crossed by some form of transgressive desire. Those of Jagatsinha, Hemcandra, Hiranmayī, Pratāp, Labangalatā come under considerable strain.

44 This is particularly poignantly brought out in Kapālkundalā's conversation with her sister-in-law:

> S. Tell me, does not the flower cause happiness by blooming?
> K. People who see it feel happiness. What is that to the flower?
> *Kapālakundalā*, *BR*, i, 109.

45 It is unfair to Rohinī and her creator to call her simplistically a scheming woman. Indeed, Rohinī's character is, in a sense, more complex and socially interesting than Kunda's simple innocence. Rohinī's evolution shows the social construction of personality, and the evolution of her sexuality is a consequence—not of something intrinsic to her, but of deprivations imposed on her by social norms.

46 Liminality and transgression have moved to the centre of modern criticism through the influence of Foucault and Derrida, and the inspiration of both is at least partly Nietzschean.

47 Cf. his KC, *BR*, ii.

48 Despite his material accomplishments, Karṇa is haunted by his dubious and undisclosable ancestry, and modern writers have seen in him a symbol of unmerited sorrow. Tagore, for instance, in his 'Karṇakuntīsaṁvad' in *Kathā O Kāhinī*.

49 Sometimes those stories are *sociologically* analysed to yield information about status of women, male domination etc. in those societies. I am concerned here with their literary and philosophical point, rather than their sociological bearing.

50 Two points ought perhaps to be made about this. It is sometimes pointed out that the materiality and references to explicit sexuality in this story are derived from earlier folk sources, because popular imagination and story-telling is less inhibited by norms of silence, taste and etiquette. Folk narratives, or folk views of the world touch without inhibition on what Bakhtin has called 'the material bodily lower stratum'. However, it should be noted that classical Sanskrit poetry has a strong, sometimes clearly dominant, element of eroticism and explicit sensuality. Bankim seems to regard eroticism, i.e. conscious and playful attention to sensual pleasures an effete high culture feature, and something that corrupts popular taste.

51 The *Gītagovinda* has the famous paradoxical double invitation to its reader: *Yadi harismarane sarasam mano*, or ironically, *yadi vilāsakālāsu kutūhalam*, in either case, the author hoped, he would find a reading of the poem rewarding, though there is also evidently a theory that the ideal reader would be a person who will appreciate both its divergent *rasas*. An ideal reader presumably would be one who would not be too devotional and thus miss the erotic pleasure, or too sensual and thus miss the latent devotion.

52 Bankim himself follows the commentary on the *Bhāgavata* by Sridharaswāmī to offer this reading of the *Vastraharaṇa* episode, in KC, *BR*, ii, 462.

53 Indeed that particular invocational *sloka* explicitly hopes that even after the metaphorical translation of these raptures a sufficient pleasurable residue is still left.

54 Wittgenstein, *Tractatus Logico Philosophicus*, 6, 43II.

55 For an account of the career of tragedy in European art, George Steiner, *The Death of Tragedy*, Faber, London, 1961.

56 Georg Lukacs, *Theory of the Novel*, Merlin Press, London, 1961.

57 *BR*, ii, 69.

Chapter 2

1 *Kamalākānta*, published first as essays in Bankim's periodical, *Bangadarśan* (*BD*) between 1873 and 1875, was brought out together as a single work in 1875. Some of the essays appearing in *Kamalākānter Daptar* were not written by Bankim. A second, revised and enlarged edition was brought out in 1885. A second edition of this larger work was brought out in 1889.

2 Bakhtin, *Rabelais and His World*.

3 Sukumār Rāy, 'Khicudi', *Āboltābol*.

4 *Bankim Racanavālī* (*BR*), ii, 49.

5 He can be called a *lumpen* character. But lumpen in marxist theory is not an independent descriptive term, it indicates a sort of degraded version of more solid social types. People have to be lumpenbourgeois, or lumpenproletariat; but it is not easy to find such a second term for him.

6 There is a simpler disjunction in the text itself. It is supposed to have been put together by a person called Bhīṣmadev Khosnavīś.

7 This is suggested clearly in the interrogation of Kamalākānta in the court scene in *Kamalākānter Jobānbandī* (*KJ*), *BR*, ii, 101ff.

8 *BR*, ii, 100.

9 I have used Naoroji as a symbol for another trend of critical thought in early nationalism. Naoroji was the author of the work, *Poverty and UnBritish Rule in India*, which enunciated the 'drain theory', and began the political economic critique of colonial rule. Yet he represents a trend of thought which thinks high capitalist principles are enactable in the colonies; Bankim on the other hand, considers that kind of re-enactment impossible. Any attempt at re-enactment must degenerate into a travesty.

10 *BR*, ii, 92.

11 *BR*, ii, 108.

12 The most erudite and convincing presentation of this argument was perhaps Bhudev Mukhopadhyay's *Sāmājik Prabandha* (Essays on Society) written between 1887 and 1889. For an account of his ideas, Raychaudhuri, Chapter 2.

13 *BR*, ii, 104.

14 This was stated much earlier in the famous essay on the babu in *LR*, through the sage Vaisampayana's omniscient speech. *BR*, ii, 11. For a discussion, see below.

15 *BR*, ii, 101.

16 *BR*, ii, 102.

17 *BR*, ii, 103–4.

18 I use the terms language, game and life form rather than the more clearly defined practice or social form, precisely because these are hazier. But a hazier term can sometimes fit our intentions more precisely, as in my view it does here. To use terms like social form or political structure would attribute more deliberation to Bankim's use than is there, and give it too clear a theoretical form. However, clearly Bankim is sensitive to the fact that court language is not an arbitrary, capricious use of language, it fits the social practice of law, and the legal structure of that kind can exist only under a colonial dispensation. It would be wrong to deny that there is a certain theoretical quality about this thinking, but equally, to extrapolate a fully structured theory of colonialism from it would be injudicious.

19 *BR*, ii, 104.

20 *BR*, ii, 105.

21 *BR*, ii, 107–8.

22 The *Gītā*, uses this idea to indicate a person free of the bonds of interest.

23 *BR*, ii, 103.

24 *BR*, ii, 9–10.

25 In this piece, and the two following pieces in *Lok Rahasya, Babu* and *Garddabha* (The Ass), the stotra form is alluded to more in terms of the semantic and rhetorical aspects, what Sanskrit rhetoricians would call in terms of the *alaṁkāras* used. Later, he experiments with a politicization of the stotra form, first somewhat satirically in *Āmār Durgostav*, *BR*, ii, 81, later more seriously, in *Ānandamath*, with enormous success.

26 Another point usually missed by admiring commentators of the Hindu religion is the great competitiveness among its deities. Usually, commentators point only to the tolerance resulting from the pantheism of Hindu religious beliefs; a worshipper of one god can nevertheless have regard for others; persons belonging to the same family can worship different gods. First, of course such possibilities should not be over-extended. In cases where ritual practices are sharply conflicting, there are real and narrow limits to such tolerance. Ritual practices of the strictly vegetarian *vaiṣnavas* can hardly be compatible with the ritual slaughter (*bali*) practised by *śāktas*. However, the intolerance is sometimes shifted to the plane of a strange form of verbal measurement. It is common to find stotras to one deity which would assert that other deities are always worshipping at his feet, or more mystically, a million brahmās would not be able to measure your end, etc. A language of excess is quite often a part of the stotra form, and Bankim sees here great possibilities of travesty.

27 *BR*, ii, 9–10.

28 *Byāghrācārya Bṛhallāngula*, *BR*, ii, 1–9.

29 *Bābu*, *BR*, ii, 10–12, which I regard as one of the best pieces of satire Bankim ever wrote. For its rhetorical elegance, political point, and the sheer fury of the emotion it expresses, it has few comparisons in the history of Bengali literature.

30 This is typical of the double meanings used in traditional writing, for the term *guna* means both virtues and the more neutral meaning, qualities.

31 For a brief account, see below.

32 This is a common phrase used by religious texts, to invigilate their own interpretation. The text itself offers alongside a sort of authoritative commentary.

33 Keśabcandra Sen, leading Brahmo reformer and leader of middle-class opinion.

34 *BR*, ii, 11–12.

35 *BR*, ii, 12.

36 Like the play on *udaracaritra* and *udāracaritra*.

37 It is on rare occasions that Tagore would indulge himself with a few, obviously non-serious examples of wordplay. The versatility of a Bengali style which uses the double meanings and convenience of *sandhi* and *samasa* is spontaneously rediscovered by Tagore in some of his witty songs, like *cā-spṛha cancala cātaka dala cala cala cala he*, *Gītabitān*, 598.

38 *BR*, ii, 92.

39 Ibid., 93.

40 Ibid., 93.

41 The original *sloka* was devoutly intended, and was meant to announce the omnipotence of the lord by whose grace even the dumb can speak, the lame cross a mountain.

42 *BR*, ii, 93.

43 Ibid., 93.

44 Ibid., 93–4.

45 Ibid., 94.

46 Ibid., 94.

47 *BR*, ii, 55–6, this piece, the third in *KD*, is called 'Utility or the Philosophy of the Stomach'.

48 *BR*, ii, 69.

49 I am not using the term to denote a narrow negative, or *opposite*. In the best of cases, it is difficult to determine precisely which social role is, in the narrow sense, the opposite of another. In the babu's case, one could say that these are his opposite

in the sense that he rules over them; but that is not equally or in the same sense true of all these types. They are called his others in the looser sense that their experience of the world was bound to be very different, and they could therefore, in satire, be seen as representatives of other principles.

50 Examples of these are: *A Criticism of the Ramayana* (written by a certain English critic), *BR*, ii, 27–29; *Letter from a 'Special' BR*, ii, 31–3, both part of *LR*; *The Big Market* (*Baḍa Bājar*), *KD*, *BR*, ii, 75–9. There are similar passages in other, more serious essays too, as in the *Kṛṣṇacaritra*.

51 *Ānandamath*, *BR*, i, 692.

52 *Devī Caudhurānī*, *BR*, i, 764–8.

53 *Indirā*.

54 *Rajanī*.

55 *Dharmaśikṣa*, in *LR*, *BR*, ii, 42–4.

56 *BR*, ii, 48.

57 *BR*, ii, 44–7.

58 *Prācīnā evaṁ navīnā* (Old Women and the New), *BR*, ii, 249–56. Interestingly, Bankim's perceptiveness about 'voices' is reflected here. After the first main essay, there are several brief comments purportedly written by women correspondents.

59 Ibid., 251.

60 Though he can describe them extremely well, cf., ibid., 252–3.

61 Walter Benjamin, *Illuminations* (trans. Harry Zohn), Fontana Collins, Glasgow, 1970, 'Thesis on the philosophy of history', 259.

62 *BR*, ii, 249.

63 *BR*, ii, 399.

64 Ibid., 399.

65 Ibid., 399.

66 Ibid., 399–400.

67 Ibid., 250.

68 *LR*, *BR*, ii, 44.

69 Ibid., 45–6.

70 *BR*, ii, 47.

71 Ibid.

72 Ibid.

73 *Prācīnā evaṁ navīnā*, *BR*, ii, 251.

74 Ibid., 252.

75 Ibid., 254.

76 Ibid., 256.

77 Mikhail Bakhtin, *Rabelais and His World* (trans. Helene Iswolsky), Indiana University Press, Bloomington, Indiana, 1984.

78 I have argued earlier that Brajeśwar in *Devī Caudhurānī*, although a protagonist placed in a historical milieu in which the babu had not arrived, is as much a babu as more straightforward cases like Nagendra or Govindalāl.

79 Tilottamā's stepmother in *Durgeśnandinī*, though her actual identity is not revealed till the end of the story.

80 *Mṛṇālinī*.

81 *Devī Caudhurānī*.

82 *Rājsinha*.

83 *Candraśekhar*.

84 This has been brought home very powerfully by the recent currency of Mikhail Bakhtin's work.

85 *BR*, ii, 373.

86 Ways in which such delicate operations can still be done, within the general ambit of marxist theory, is shown in Volosinov, *Marxism and the Philosophy of Language*, (trans. Ladislav Matejka and I.R. Titunik) Harvard University Press, Cambridge, Mass., 1986. I simply invoke this text in favour of the view that marxism may

have something to do with intangible but fundamental things like words and language too, besides less in tangible ones like production. It is not claimed however that the analysis undertaken is an application of the theoretical insights of Volosinov.

87 The linguistic expert in Shaw's *Pygmalion.*

88 This is especially clear in his occasional references to the state of Bengali literature, and in the introductory remarks to the readers of his journal *Bangadarśan.*

89 He advocates this theory explicitly in his essay 'Bāngālā Bhāṣā' (The Bengali Language), *BR*, ii, 368.

90 Notice the practically identical description of the magistrate's court in *Mucirām Guḍer Jīvancarit*, *BR*, ii, 119.

91 *BR*, ii, 112.

92 *BR*, ii, 112.

Chapter 3

1 This chapter discusses Bankim's major theoretical work *Kṛṣṇacaritra* (*KC*). In my view this is a theoretical more than a historical work; for its 'historical' arguments could easily be faulted even by Rabindranath Tagore, a person not known for his theoretical acuity, cf., his '*Kṛṣṇacaritra*' in *Ādhunik Sāhitya*. Bankim's views on religion and social theory are scattered over all his works, both fictional and ratiocinative. But they are more particularly grouped together in three major works, *Kṛṣṇacarita*, *Dharmatattva* and *Srimadbhagavat-gītā* which is an unfinished commentary up to Chapter Four (*jnānayoga*), sloka 19. Besides, in many short articles Bankim interpreted Hinduism indefatigably, cf., for instance, '*Devatattva o Hindudharma*'. And it is not surprising that some of his most significant thoughts about how to behave correctly in different situations are to be found in his novels. As this paper concentrates on *Kṛṣṇacaritra* it simply outlines a part of what Bankim had to say on the matter of religion.

2 Ludwig Feuerbach, *Essence of Christianity*, Ungar, New York, 1957, introduction.

3 The term classical is meant in the sense in which art criticism uses it.

4 Indeed, Bankim in my view carries on a more or less explicit polemic with all these texts which between them constitute and structure the canonical vaiṣṇava tradition.

5 This is a hermeneutic question that arises out of Bankim's argument, but one he chooses to ignore. If we show how different the figure of Kṛṣṇa is in the successive texts, yet it is evident that they are not perplexed or unduly troubled by such inconsistency, obviously there appears a question—why do they not think so? On the simplest hermeneutic principle that if we find something unacceptable and wrong, we must enquire why others did not think as we do. This could be for several reasons. It could be that their criteria of rightness and ours are the same but they made a mistake in this case, or that their criteria are quite different from ours.

6 This is the way of figure the Rādhā is first interpolated, and gradually comes to dominate the story both theologically and aesthetically.

7 For instance, deeply religious texts like gospels and acts within the first order texts of Christianity and the works of St Augustine and Thomas Acquinas are all solidly attributed to individual authors, in contrast to either the celebrated authorlessness of some texts like the *Vedas* or the *Gītā* (which has an extremely interesting status on the question of authorship), or the flimsy attributions of the *Mahābhārata* and the *Purāṇas.*

8 At least partly this is the reason behind elaborate rituals of replacing intentions by a theory of appointment. Authors do not say that they decided or chose or intended

to write a book—which requires a special individuality. Rather they are appointed by someone to do this, and consequently act as something between an instrument and an amenuensis. Behind the elaborate convention of invocations this may be the real operative point.

9 It may not be accidental that the term for this is so metaphorically appropriate—*sūtra*, a thread. Here it is a thread which binds ideas constituting a complex thought with one another; it also binds or joins performances, forming a connection between different individual minds which are thus threaded together.

10 For instance it is often said in the epics and the *Purāṇas* that an individual is born as 'a part of' some god (*anśa*), which presumably should be read as his representing only a part or a fraction of the deity's powers.

11 In European philosophy this is a central problem, and is commonly given two well-known answers. One solution is to say that the infinite cannot be known in the way in which ordinary objects can, and must be the object of faith. Cf. Kant, *Critique of Pure Reason*, Dent, London, 303ff., 342ff., or in Pascal's *Pensee*. A second solution, found in Hegel, for instance, holds that if there is any minimal order in this infinity, this infinity is knowable.

12 Hiraṇyakaśipu had, on the common version of the story, earlier asked for immortality, and Brahmā refused claiming some procedural difficulties in conferring a boon of such magnitude. Hiraṇyakaśipu, who seems to have fancied himself as a logician, then asked that he should not be destroyed during day or night, by man or beast or god, either inside a house or outside, neither by hands nor by weapons. Obviously now he considered himself fully insured and to have logically covered all possibilities. But clearly he had not reckoned with the omnipotence of God, and Hiraṇyakaśipu's friends could add, with His mendacity. For the demon had correctly thought that no further types could be invented—i.e., he thought that if x and y were the only natural possibilities, his boon meant that he could only be killed by z. God obviously interpreted this to mean half-x half-y which gets round the problem rather unconvincingly, and is evidently a foul. The argument unfortunately is invalid despite its unimpeachable source. Thus, although we must rejoice at the demise of a person causing great evil, our logical sympathies must lie with the late demon. Sugata Kaviraj pointed out to me the unacceptability of the divine interpretation.

13 The standard sequence of the ten *avatāras* of Viṣṇu, are the *matsya* (fish), *kūrma* (turtle), *varāha* (boar), *nṛsiṁha* (man-lion), *vāmana* (dwarf) and the anthropomorphic series of Paraśurāma, Rāma Balarāma, Kṛṣṇa and Kalki. Jayadeva, with real genius, supplants Kṛṣṇa by the Buddha, obviously on the ground that Kṛṣṇa is Hari himself, and it is travesty of sorts to say that he is an *avatāra*. As a result the powerful appeal of the Buddha is enlisted in the service of the vaiṣṇava religion. Cf., Jayadeva, *Gītagovinda*, canto I, sāmodadāmodarah, sloka 13. He is quite consistent about his alternation of the daśāvatāra, for a little later he congratulates Kesava for spreading compassion (*Kāruṇyamātanvate*), I, 16.

14 *Gītagovinda*, canto I, 13.

15 Generally known as the gaudīya vaiṣṇava tradition.

16 For a learned account of this, Sisir Kumar Das, *The Mad Lover, Essays on Medieval Indian Poetry*, Papyrus, Calcutta, 1984; especially, 'Radha: A Mad Lover', 69ff. For a more detailed examination, Sasi Bhusan Dasgupta, *Srīradhār Kramavikās* (in Bengali).

17 Rabindranath, with instinctive poetic sympathy, celebrates precisely this aspect in his 'Vaiṣnav Kabitā', *Sonār Tarī*, *RR*, vol. i, WB Govt., 1961, 367.

18 *KC*, *BR*, ii, 191, 464.

19 Although he does not use the term 'classicization', Sudhir Kakar makes a similar point about the *Gītagovinda*. Sudhir Kakar, 'The secret passion of Radha and Krishna' in Veena Das (ed.), *The World and the World*, New Delhi: Sage Publications, 1986.

20 For those with an uncompromisingly empirical bent of mind, this could be a report of what actually happens, embellished by a great deal of simply verbal luxuriance

about sex. Rabindranath would use his powerful distinction and say that it is short on facts but rich in truth.

21 The technical name of the idea central to that doctrine is *premabhakti*.

22 I have argued later, in the next chapter, that an identical antithesis is observed in the character of Sītārām—when he acts on the basis of reason he is capable of great historic action; when he falls into sensuality, through his attachment to Śrī, he becomes gradually incapable of historical or practical initiative.

23 As we shall see, Tagore has a very different *theory* of the significance of the Kṛṣṇa figure which he sets out with clarity and eloquence in his critical appraisal of Bankim's *KC*. On his view, Kṛṣṇa's story is poetic and therefore aesthetically true. It is doubtful if Tagore would have treated Kṛṣṇa's life as a practical text in the same way as Bankim; in any case he would not regard it as a text created by a historical Kṛṣṇa, but by a poet, narrator, who tried to fix poetically the meaning of being an excellent, complete human being—*puruṣottama*. For an interesting recent account of the difference,—why Rāma is *naracandramā* and Kṛṣṇa *puruṣottama*, see B.K. Matilal, 'Ram Naracandramā, Kṛṣṇa Puruṣottam, *Deś* (in Bengali), 7 June 1986.

24 As Trivedi puts it in his address, that image is not compatible with *yugadharma*—the ethic of this historical period, and what it requires, *Carit-Kathā*.

25 *KR*, *BR*, ii, 407. Not surprisingly, the first chapter is called 'the purpose of the book'.

26 Ibid., 408.

27 'Master narrative' is a concept used normally and widely in cultural and socio-psychological analysis.

28 This is close to what Tagore called *satya* or poetic truth.

29 Ludwig Feuerbach, *The Essence of Christianity*, edited and abridged by E.G. Waring and F.W. Strothmann, F., New York: Ungar Publishing Co., 1957, Chapter II.

30 *KC*, *BR*, ii, 408.

31 Ibid., 408.

32 Tagore argues this very forcefully against Bankim in his *Ādhunik-Sāhitya*, Viśwabhāratī, 1966, 69ff, especially 80–3.

33 This could be one reading of the fact that little is reported from his earlier life. Another view could be that it is the effectiveness of the context: even though he had a colourful adolescence, it was not strictly relevant to the story of the Mahābhārata war.

34 *KC*, *BR*, iff, 409.

35 *KC*, 410.

36 *KC*, 412

37 Tagore's theory of a poetic truth, if reconstructed systematically, would appear very similar to Hans Georg Gadamer's *Truth and Method*.

38 *KC*, 427–8.

39 Rabindranath Tagore, '*Bhāsā O Chanda*'.

On this point Tagore showed complete consistency. When he argued seriously in his article on Bankim's book, he takes the same line:

> 'In such cases the conjectures of a poet could come much closer to real historical truth than the conjectures of a historian . . . .
>
> Facts, which are called facts in English, truth is far larger than that. One has to extricate truth from the heap of facts by reasoning and by imagination. Quite often we get heaps of dry faggots of information in history. But truth is evoked in poetry by the poetic genius . . .
>
> If Kṛṣṇa had a history, that would perhaps have contained a thousand incidents which would have no permanent value even though performed by Kṛṣṇa himself. That means that these acts would not have expressed the Kṛṣṇaness of Kṛṣṇa. Besides, since it is impossible to know everything, some of these may have appeared to be even contrary to his proper nature. Every person does a great number of things which are inconsistent with what he really is'.

40 *KC*, Chapter 10, 427–9, *Prakšiptanirvācan*
41 *KC*, 428–9.
42 '*aghatanaghatanakaušala*', *KC*, 429.
43 *KC*, 429.
44 For it is in the *Gītā* that Kṛṣṇa is rather strongly inclined towards a form of divine exhibitionism. Besides the episode of the *Viśwarūpa*, which could be called infinitely exhibitionistic, on innumerable occasions in the *Gītā* he asserts his equivalence with the infinite: and the whole sermon is given in the presumptuous (some would say, high) first person. For a person who could say things like 'sarva dharmān parityajya māmekam śaraṇam braja', etc. may be gifted with infinite amounts of all other commendable qualities, but not modesty. For Bankim this ought to create an obvious problem of logical consistency. If the mark of the first layer of the epic is that Kṛṣṇa's godliness is not declared, and that is what makes it valuable and rationally acceptable, this clearly goes against the *Gītā*, clearly his favourite among religious texts. The *Gītā* announces His godliness blatantly and constantly, and it also takes recourse to the supernatural to compel belief, since the *viśwarūpa darśaa* is not reduced materialistically or rationally to a possible hallucination of Arjuna. Although this is crucially significant to his case, Bankim does not give this an elaborate treatment claiming that he would do it elsewhere. This presumably is his commentary on the *Gītā* which does not actually come upto this episode.
45 *KC*, 429.
46 Rajśekhar Basu, *Mahābhārater Sāranubad* (in Bengali), introduction, M.C. Sarkar and Sons, Calcutta, 1356 (Bengali era).
47 *KC*, 431.
48 *KC*, 431.
49 *KC*, 432.
50 *KC*, 432.
51 *KC*, 432.
52 Bankim does not use another possible argument here—that this theory on top of being wrong is also presumptuous, because it thinks that it is our responsibility to save Him from suffering. We cannot do this, because He has more powers than we have, also because He has more knowledge than we have. It could illustrate what could be called, misquoting Marx, the world-consuming pretensions of the babu, for he typically assumes that he is a trifle more powerful than the Almighty and a shade more clear-headed than Him. Bankim does not employ the argument for reasons of economy, but I am sure he would have if he saw an occasion for making an additional dig at the babu.
53 The first argument is more powerful than the second. Argument B says that even if God suffers it must be for an instant. But A is more radical because it shows He does not suffer, or to speculate whether He suffers is a mispredication, and therefore the question of whether for an instant or longer and the comparability of His with our time does not arise. However the idea of a scale-relativity to time is intriguing. He asserts that the relation between God and men is like the size difference between Gulliver and the Lilliputians etc. Relativity of time can be constructed in several ways—this way of doing it is rather odd.
54 *KC*, 433.
55 The form of this treatise is interesting; it is written as a *dialogue*, a form that enjoyed great popularity in the Enlightenment tradition. Some of its purely discursive characteristics ought to be remarked. It does not present the preferred ideas with a false impression of overwhelming rationality, which implies that anyone who denies them is deficient in rational thought. On the contrary the dialogue form grants a certain rationality to the interlocutor, because he is allowed to talk back and argue, which implies at least that there is a possibility of an argument on the opposite side. Rationalist authors often show a strong preference for a dialogical form in which in the course of stating the grounds of belief on both sides, the other view is undermined. Early Christian missionaries composed some religious treatises in which Hinduism was seen arguing against some form of Christianity,

to the detriment of the Hindus of course. Thus Bankim's book could perhaps be seen as an answer in kind. But this was a performance in which, typically, he uses techniques of rationalist culture to undermine its theory.

56 *KC*, 434. Bankim calls these kind of people, who hold strictly Cartesian views, 'Seśwara Vaignānika', theist scientists.

57 The terms 'natural' and 'function' are obviously evolutionist.

58 *KC*, 434. The equanimity with which Bankim introduces this doctrinal change is really astounding. For the first three laws of the universe are simply the three functions of Hindu theology under the flimsy positivistic garb of 'laws'—creation, maintenance and destruction. Bankim simply adds the Spencerian idea of improvement as if nothing is more natural.

59 Partha Chatterjee, 'Culture and power in the thought of Bankimchandra' in Thomas Pantham and Kenneth Deutsch (eds), *Political Thought in Modern India*, New Delhi: Sage Publications, 1986.

60 On the point of Comtean sociology, Jasodhara Bagchi's work is perceptive.

61 That seems to be the main point in the presentation of his character in the *Caitanyacaritāmṛta*.

62 Namwar Singh, *Dusrī Paramparā Kī Khoj* (in Hindi), Rajkamal Prakashan, Delhi, 1983. Singh follows up the search begun by Pandit Hazārī Prasād Dwivedī, one of the finest and most original of Sanskritists of his generation. Singh's book too makes a highly original point by claiming that radicals can have two possible attitudes to their tradition. They can convince themselves that they are the first nonconformist or critical people in history, and the whole earlier antiquity was oppressive, dishonest, cowardly, in which case, they face a wholly monolithic tradition which has to be entirely destroyed. Alternatively, they can believe with Engels that there must have been nonconformist critical people before them, and these trends have to be rediscovered and respected. Singh, if I have understood him correctly, eloquently defends the second view.

63 *Caitanyacaritāmṛta* (*CC*), I, V.

64 *CC*, I, vi.

65 See the *CC* passim. Especially the slokas which are recommended as his particular favourities; eg. the splendid sloka in *Madhyalīlā*, I, vii. In the translations appended to the edition by Sukumar Sen, strangely, the slokas are often incomprehensibly de-eroticized. For instance, this particular sloka from *Kāvyaprakāśa* contains a phrase '*kaumāraharah*' which is wanly translated as 'the person who stole my mind' rather than 'the person who stole my virginity', a somewhat different proposition.

66 Tagore, 'Vidyāpatir Rādhikā', *Ādhunik Sāhitya*, 63ff. However the variety that Tagore notices and interprets so beautifully are all internal to the figure of Rādhā.

67 Interestingly, the ability to accomplish the impossible and the excessive is a general characteristic of popular heroes—from the popular Kṛṣṇa to Rabelais's Gargantua. Cf. Mikhail Bakhtin, *Rabelais and His World* (trans. Helene Iswolsky), Bloomington: Indian University Press, 1984, Ch. 5. Bankim is uneasy, occasionally contemptuous, about the story of sixteen thousand consorts and one lakh eighty thousand 'sons', which are essentially examples of a popular doctrine of excess or a loving approximation of infinity.

68 *KC*, 454. Sometimes the slide into eroticism is given a longer history, with Jayadeva seen as the decisive and irreversible turning point.

69 For an interesting contemporary example, see Ramchandra Gandhi, *I Am Thou: Meditations on the Truth of India*, Poona: Indian Philosophical Quarterly Publications, 1984. Gandhi's position raises interesting paradoxes for both those who would wish to support or oppose his kind of interpretation. On the one hand, it raises problems of anachronism: for the view that religious ideas are systematically metaphorical appears to me to be unmistakably modern, a result of accepting a disenchanted world, a natural world of causal processes rather than a world of meanings. At the same time, to claim that all these were literal beliefs also seems hideously condescending as metaphorical ideas must have constituted some part of the traditional religious system of beliefs. To determine exactly what was

regarded as metaphorical and what an unmetaphorical statement thus becomes a very difficult task.

70 'The mad lover' and 'Radha: a mad lover' in Sisir Kumar Das, *The Mad Lover*, Papyrus, Calcutta, 1984.

71 The *CC* is full of admonitions from Caitanya on this point, for example, in the story about Sanātana, *CC*, 300ff.

72 *KC*, 462.

73 L. Kolakowski, *Religion*, Fontana, Glasgow, 1982.

74 *Dharmatattva*, *BR*, ii, 588.

75 Characteristically, Bankim refuses to provide a Bengali translation of the offending sloka in the *Vastraharana* episode in his own discussion, because of what he considers obvious sexual perversity (rather than explicit references to the body). Kakar emphasizes the fantasizing streak in the *Gītagovinda*, Cf. Kakar.

76 *KC*, 464.

77 *KC*, 466.

78 For instance, the theory of *premabhakti* in the *CC* sees Rādhā as an alienation of God's own self into another serving and observed self, a sort of arrangement of infinitely extended narcissism. Rādhā is the self of Kṛṣṇa alienated into an other; through her He can see Himself, and in her or seeing her, He can also see Himself being seen. *CC* does not set out this theory in doctrinal detail, except in the very laconic opening slokas, but it is implicit in its more metaphysical parts.

79 *CC*, *Madhyalīlā*, VII, 38.

80 A certain worry in the popular mind persists even today about this inconsistency. On the one hand Kṛṣṇa is the lawgiver to the entire universe; he is an avatāra of that great divine self of his. In his human incarnation however he is undoubtedly a law-breaker, particularly of a rather crucial norm of domesticity. Popular consciousness could understandably be troubled about this paradox, although tempted to solve all such puzzles by the *deux ex machina* that everything that did not square up was His līlā, that divine justice often appeared to limited human minds as injustice; that this did not indicate the unjustness of the action, but the limitedness of our minds.

To illustrate this let me quote a very ordinary Bengali folk song of the fifties. It is not distinguished for either its literary value, or its tune, or its humour. But it places the popular dilemma at the centre.

> *lālpāgḍi bendhe māthe*
> *rājā hale mathurāte*
> *bānśi cheḍe danda hāte*
> *bandhu hale dandadātā.*
> *ekhan kalankinī rādhāy danda*
> *nā dile mān thāke kothā?.*

Translated blandly this means;

> You go to Mathura and become the king.
> you put on the red turban on your head,
> you exchange the flute in your hand for
> the royal sceptre; from the lover you
> become the dispenser of justice; now
> if you do not punish the transgressions
> of Rādhā (or Rādhā the sinful
> woman) how can you save your honour?

This illustrates the problems: God cannot transgress the rules He is supposed to uphold; He cannot Himself do what He is supposed to punish others for doing. The song covers the main problem by politeness; for it neglects to mention the accomplice of Rādhā's transgressions. However, the stanza is fascinating for another sociological reason: it is replete with an amazing modern symbolism. Two symbols of power are used; one is the unsurprising sceptre which both the king and the ruler of the universe can conceivably hold, to underline their eminence. But the other symbol is intriguing. *Lāl pāgdi*, the red turban, traditionally associated with zamindars' minions, was worn by the Calcutta police constables, from the British days, and they were often referred to irreverently as simply *lāl pāgdi*. What the song fascinatingly does is to simply enlarge the symbol of power in terms of scale. Red turbans are indicators of power not only in Calcutta streets, but also in ancient Mathura, and by extension in the universe in general. I find this idea of the almighty sharing some characteristics with the Calcutta constable irresistible. It is not at all illogical and may indicate a liberal conception of creation. For He is after all the constable of the universe and shows the same mixture of omnipotence and superfluity. He does essentially the same thing though His beat is somewhat more extensive. Since then, alas, Calcutta policemen have been divested of this article of clothing, and consequently, of their similarity with the Almighty.

81 Though this is more so in the case of Rādhā than say Sītā and Draupadī, the two other feminine ideals in the Hindu tradition. But Rādhā's character is double-edged and transgressive. It is in fact difficult to press her into service as an unproblematic ideal; obviously she could not be an ideal wife, and the idea of an ideal mistress (occasionally evoked in some vaiṣṇava doctrines) is itself deeply ironical in the moral universe of the Hindus.

82 *CC, Madhyalīlā*, I, 7.

83 *Māthur* is the popular name used in *Kīrtan* cycles of that part of the story in which Kṛṣṇa goes away to Mathura, and under the pressure of unspecified business, becomes increasingly distant from his friends in Vṛndāvana.

84 *Meghadūta*, 21 fn in the *Sahitya Academy*, edition, 1982, regarded as authoritative. My point is not the authenticity of the verse, but the conceptual point it makes.

85 Tagore, *Bhānusinher Padāvalī*, *RR* (Collected Works), I, W.B. Govt. 1961, 142.

86 Ibid., 142, the famous poem, *Maraṇa*; 'maraṇa re tunhu mama śyāma samān'. In later poems too this equation of death with love/marriage/union is carried on, as in the celebrated poem: *jhulan*: '*āmi parāṇer sāthe kheliba ājike maraṇakhelā/ niśīthabelā* . . .'.

87 When Mahendra, one of the principal protagonists in *Ānandamath* is taken into the underground temple he finds three figures of the mother—what she was, what she is, and what she will be. Her worshippers however are vaisnavas, an extraordinary state of affairs. From this point of view this is a simple substitution of *Rādhā* by *Śakti*.

88 *BR*, ii, 269.

89 *KC*, *BR*, ii, 581.

90 *DT*, *BR*, ii, 647–60, gives the argument in full.

91 *KC*, 612.

92 The fuller argument is set out in the next chapter. Here misleading because vaiṣṇavism is normally associated with non-violence and does not fit the advocacy of war-like solutions. After all, the *CC* advises vaiṣṇavas to tolerate and forbear: *tṛṇādapi sunīcena taroriva sahiṣṇunā* etc.

93 Reinforced by his unfinished commentary on the *Gītā*, *BR*, ii.

94 Bankim sometimes uses this argument as well.

95 Partha Chatterjee, op. cit.

96 *Kamalākānter Patra*, *BR*, ii, 92.

97 *DT*, *BR*, ii, 618.

98 Victor Turner, *Dramas, Fields and Metaphors*, Ithaca: Cornell University Press, 1974.

99 *Gītā*, canto IV, 8.

Translations of slokas cited in Chapter 3

14. Moved by deep compassion, you condemn the Vedic way that ordains animal slaughter in rites of sacrifice. You take form as the enlightened Buddha, Kṛṣṇa. Triumph, Hari, Lord of the World. (Barbara Stoller Miller, *Gītagovinda of Jayadeva*, Delhi: Motilal Banarsidass, 1984, 7, I.

23. Miller translates this more poetically as 'until the secret passion of Rādhā and Mādhava triumph on the Jamuna—riverbank' (69). But it could also be more literally translated as 'let Rādhā and Mādhava triumph who thus play on the banks of the Jamuna'.

35. It is a narrative of old times, which contains instruction about *dharma* (religion/ethics), *artha* (material affairs), *kama* (love) and *mokṣa* (salvation), that is called history.

73. The transformations of the love of Rādhā and Kṛṣṇa is the power of infinite love. Once they assumed different forms on this earth even though they are indeed indivisible and identical. Now those two different forms have been reunited in the joyous figure of Caitanya. Let us worship the Lord who has the external manifestations of Radha but is Krsna's self.

74. The full *sloka* is:
How great is the love of Rādhā for me; how indeed is my own *mādhurya* which she tastes, and how exactly does she feel when she favours my sweetness—in order to know this Hari, filled with Rādhā's attributes (bhāva) was conceived in Śacī's womb.

89. The person who is called *dhīralalita* must be cultivated, youthful, humorous, untroubled, and would often be under the influence of the woman he loves.

91. The person who took my virginity when I was young now wants me; this night is like that first night of Caitra; the wind carries the fragrance of the same flowers; I am still the same me; we can still enjoy our union. Still, why does my soul wish to go back (attain again) our (first) union in the arbour on the banks of the Rebā?

92. Let your foot quell the harsh sun
Burning its fiery form in me to torment Love. (Miller)

93. You are my ornament, my life,
My jewel in the sea of existence.
(Miller, 112).

Chapter 4

1 I do not mean to deny that history is only one among many possible ways of setting up an intelligent relation with the past. Myths and other narratives often accomplish a similar task. It is hazardous to have a purely chauvinistic temporal preference in favour of history.

2 They do not seem to ponder the philosophical paradox of past existential statements: while their existence in an empty sense is not doubted, they depend peculiarly on accounts. There is no entry into any historical incident except through the immense vulnerability of accounts which tell about them.

3 For a detailed and perceptive account of two such views, Tapan Raychaudhuri, *Europe Reconsidered*, Oxford University Press, Delhi, 1988.

4 Ibid.

5 No doubt this way of treating all European writing about Indian history as an entirely homogeneous, undifferentiated whole, was unjustified. There were significant differences among European writers. But the seeing of this homogeneity was both true and false. It was true in the sense that it saw these divisions as minor elements of variation in a powerfully single ideological grid. Since Indian writings were ideological as well this was bound to happen, because ideology always produces such monstrous homogeneous evil adversaries for easier prosecution of their own case.

6 *BR*, ii, 235.

7 Assertions to this general effect are too numerous in Bankim to be listed. I have referred only to the more colourful of his statements on this matter.

8 Bankim is indeed a pioneer in this—seeing history writing as a narrativizing act, as a *muthos*, a theoretical function quite different from the basic task of fact gathering.

9 *Swapnalabdha Bhāratvarṣer Itihās* was the title of Bhudev Mukhopadhyay's famous book, a boring and unimaginative work, except to historians and sociologists. Clearly it takes after the constructive principles of famous utopias like Moore's.

10 In a quite unmysterious and unrhetorical sense. Here I have simply followed Arthur Danto's argument about the nature of narrative sentences. By changing the narrative zero, we can alter the values of all items in the narrative series. A.C. Danto, *Analytical Philosophy of History*, new edition published as *Narration and Knowledge*, Columbia University Press, New York, 1985.

11 Ernest Gellner uses this characterization about much of the histories that nationalist movements gave to their respective nations. Gellner, *Nations and Nationalisms*, Blackwell, Oxford, 1981.

12 For a different view, which accepts the point that Gellner indicated by 'fraudulence', but considers it inevitable and therefore a normative condemnation rather pointless, Benedict Anderson, *Imagined Communities*, NLB, London, 1983.

13 After the title of Cornelius Castoriadis's well-known book.

14 Bankim points to this sharply in a brief but bitter parenthesis in *Kamalākānta* which we have read earlier in Chapter 2, *BR*, ii, 92.

15 *BR*, ii, 97.

16 Edward Said, *Orientalism*, Routledge and Kegan Paul, London, 1978.

17 Laboriously academic constructions of the past were done by scholars like Rakhaldas Bandyopadhyay and more famously by Jadunath Sarkar. It is not entirely surprising that Sarkar was asked to write introductions commending the historical veracity of Bankimchandra's novels.

18 Bankim began and left unfinished several fragments on the history of the Bengalis which began, as was common in his times, with anthropological reconstructions of early history and origins of language.

19 R.C. Dutt wrote two celebrated historical novels somewhat in the Bankim mould, though entirely without similar complexity, richness or humour, *Rājput Jivan Sandhyā* and *Mahārāṣtra Jivan Prabhāt*.

20 Rakhaldas Bandyopadhyay wrote historical novels on the hazily glorious period in Bengali history, like *Mayūkh*, *Dharmapāl* and *Pāṣāṇer Kathā*.
21 *BR*, ii, 234 (about India), 336 (about Bengalis).
22 *BR*, ii, 333–4.
23 *BR*, ii, 234–44.
24 Bankim is partly conscious of the many-valued meaning of the term, *BR*, ii, 240. For an interesting discussion which starts off from Bankim's remarks, Asok Mitra, 'Sat koti kotir itikatha', *Vibhāv*, Puja, 1987.
25 I must make clear what I mean by fuzziness. Those who use this way of acting do not necessarily find it difficult to characterize individuals or situations clearly. Usually they have elaborate procedures which can take off only on a clear identification of the case. But there is no identification which is overriding all others, and secondly, these communities do not feel an urge to enumerate themselves. Arithmetic calculations are not at any rate preconditions for religious rites.
26 My argument in this part depends on some suggestions made in Partha Chatterjee's discussion on colonial discourse, though my substantive differences with him would be quite evident. I am particularly indebted to him for allowing me to read his work in manuscript.
27 *Nationalist Thought and the Colonial World*, Zed Press, London, 1986.
28 This move is hard to characterize: is it compliance or inversion?
29 As an intelligent user of language and concepts Bankim discusses the conceptual difference between dharma and religion on several occasions. E.g. *Dharmatattva*, Chapter 3, *BR*, ii, 589, and Appendix B, 672. Dharma is not any religion; but what is common to them all is the nature of man.
30 And certainly not coterminous with 'material'.
31 *BR*, ii, 92.
32 Bankim, in coarsening anger, asks occasionally what the ancestors of the British were doing when Indians were composing their early philosophical texts. Actually the line of descent of the nineteenth-century Bengali babu from the composers of the *Vedas* was of course as unquestionable as that from ancient Greeks to Victorian England. Such small matters are apt to be forgotten in the heat of civilizational contests.
33 Max Müller, *Sacred Texts of the East*.
34 It was common for educated people of those times to compare with entire seriousness characters from Kālidāsa and Shakespeare, and come to quantitative measures of greatness. Bankim did this in his famous essay, 'Śakuntalā, Miranda O Desdemona'; though his finest criticism can be found, in my view, in the long unjustly neglected essay on Bhavabhūti's *Uttararāmacarita*.
35 'Construction of otherness in Marx and Weber' Occasional Paper No. 53, Nehru Memorial Museum and Library, April, 1988.
36 Tagore, 'Aitihāsik Citra', *Itihās*.

> This hunger for history is just a seed. We now realise that the Congress has not rained a series of fruitless appeals to the seat of power every year. It has sown a seed of consciousness in our minds drawing the different provinces of India closer together. Today just as we wish to be close to Bombay, Madras and Punjab, we wish as much to see the India of the past.

37 *Sāmājik Prabandha*, cf., Raychaudhuri, *Europe Reconsidered*.
38 Tagore, 'Bhāratvarser Itihās', *Itihās*, 4.
39 Ibid.
40 Ibid.
41 On this point Bankim's argument and Tagore's are almost identical. Only the tone is different, Bankim's cruel and sardonic, Tagore's hurt and sentimental.
42 Bankim does not forget to mention that they have also made a modest sum of money by writing their accounts. *BR*, ii, 336.
43 Tagore, *Itihās*, 1, 2.

44 Tagore is a particularly eloquent representative of this theory, cf. the first two essays in *Itihās*.

45 Ibid.

46 *Bhāratvarṣe itihāser dhārā*, ibid.

47 For a very similar passage, *BR*, ii, 342. Tagore's justification appears in a review article in *Itihās*, 117.

> To determine what is true in history and what is false is a very difficult task. Words grow on like animate beings; they go on changing from account to account, from one time to another. Invention is a natural faculty of the human mind—whatever account or incident comes its way it cannot keep that from being transformed, but always recasts it by pouring it into its own mould. That is why everyday we get new accounts of the same event from new people . . . Past states of affairs are not determined by events alone; how these events appeared to people is also a major object of enquiry. Therefore in the narration of historical events the way the human mind mingles with objective incidents—that is historical truth.

48 As a result of the work of the Konstanz school, especially of Hans Robert Jauss and Wolfgang Iser. Reception aesthetics gives rise to some very difficult problems, and my own material and arguments would have to be cast in a different form if I were to attempt a consistent application of the reception theory. I wish merely to state that here is excellent ground for application of the reception *problematique*.

49 A point clearly perceived by Tagore in the remarks quoted earlier—the felt nearness to people of Punjab, Madras and Bombay was intimately connected with the will to feel a nearness, community with the past, with the Aryans in this case. By implication, both these communities—of the present and the past—are imaginary, or rather these are part of the same process of construction.

50 I have discussed this in an earlier section of this work, cf. Chapter 2.

51 *Sāmya* (Equality), *BR*, ii, 381–406.

52 *Bangadeśer Kṛṣak*, (The peasantry of Bengal), *BR*, ii, 287–314.

53 *BR*, ii, 336.

54 *BR*, ii, 330.

55 M.N. Roy, *India in Transition*, Minerva Associates, Calcutta.

56 *Bāngālīr Utpatti* (Origin of the Bengalis), *BR*, ii, 344ff.

57 *Bāngālār Itihāser Bhagnaṅśa.* (Fragment on the History of Bengal), *BR*, ii, 340.

58 *BR*, ii, 337.

59 *BR*, ii, 337.

60 *BR*, ii, 337.

61 Compare for instance the framed opening passage of Tagore's essay on Indian history with the following:

> But do we find any historical material about Bengal in all this? In our view not a single English book contains the real history of Bengal. All that these contain are reports about those Muslim rulers who decked themselves with meaningless titles of nawabs and subedars of Bengal, lay prone and unworried in their beds. These report the births, deaths, factional squabbles and picnics. That is not the history of Bengal not even a part of it. It bears no relation with the history of Bengal. It contains no history of the Bengali people. *BR*, ii, 336.

62 Despite occasional unease, Bankim's treatment of Muslim rulers is generally hostile. Individual Muslim characters are often treated with considerable sympathy, occasionally great artistry, e.g., Āyeṣā, Mubarak, Zebunnisa (whom Tagore calls the real tragic heroine of *Rājsinha*). But he considers India having been in subjection for seven hundred years. His discomfort with this line of thinking is however quite real. Cf. the postscript to *Rājsinha* and the interesting discussion on freedom and subjection of India, *BR*, ii, 241ff. Since a form of uninformed secularism is currently in fashion (which believes that the more uninformed you are about the Indian

tradition, the more secular you are) and it sees the conflict between modern secular and communal forces as stretching endlessly into history, and classifies past writers into ancestors of modern political parties, it is perhaps useful to quote that postscript. Since, however, such judgements are based on rumours about Bankim rather than a reading of his texts, it is unlikely to have much effect; that is the power of political rumours. However, here is what Bankim says:

> The author wants to make the humble submission that no reader should think that the purpose of this book is to indicate some sort of difference between Hindus and Muslims. If someone is a Hindu he is not automatically good, and a Muslim evil; or a Hindu automatically evil and a Muslim good. Good and evil exist in each in equal measure. Indeed, it must be acknowledged that since the Muslims were the rulers of India for so many centuries, they must have been better in statesmanly qualities than contemporary Hindus, in others, Hindus abler than Muslims. Along with other qualities he who had the quality of dharma on his side, whether a Hindu or a Muslim, was superior. Even with other qualities, he who lacked dharma, a Hindu or a Muslim, was inferior. *BR*, ii, 661.

63 *BR*, ii, 241ff.
64 *BR*, ii, 234ff.
65 In *Religion of India.*
66 *BR*, ii, 328–39.
67 *BR*, ii, 239.
68 *BR*, ii, 240.
69 *BR*, ii, 240–1.
70 *BR*, ii, 241.
71 *BR*, ii, 214ff.
72 *BR*, ii.
73 *Europe Reconsidered.*
74 *Haraprasād Śastrī Racanasangraha* (ed. Satyajit Chaudhuri).
75 *Akbar Saher Khos Roz*, *BR*, ii, 953.
76 *BR*, ii.
77 Ibid.
78 Ibid.
79 Ibid.
80 Ibid.
81 Qazi Nazrul Islam, Kaṇḍārī Hunśiār.
82 Preface to the fourth edition of *Rājsinha*, *BR*, i, editor's introduction, p. 40. 'I have never written a historical novel before. *Durgeśnandinī*, *Candraśekhar* or *Sītārām* cannot be regarded as historical novels'.
83 The two great examples are Śānti and Praphulla.
84 Lukacs shows this at great length in *The Historical Novel.*
85 In Dickens's *A Tale of Two Cities.*
86 *Ivanhoe.*
87 Alessandro Manzoni, *On The Historical Novel* (tr. Sandra Bermann) University of Nebraska Press, London, 1984.
88 Barthes called Voltaire 'the last happy writer'.
89 *Āmār Durgotsav*, *BR*, ii, 791.
90 Haraprasad Sastri, his great admirer, recounts the discomfort most of his literary friends evinced on hearing the song. It did not seem right by any formal criteria. Bankim, Sastri reports, was entirely unmoved by such formal criticisms and trusted his instinct that it was nevertheless right.
91 *BR*, ii.
92 *BR*, i, 673.
93 *BR*, i, 676–7.

94 i.e., exactly what the hermit says to Mahendra in their conversation.
95 Cf. T. Raychaudhuri, *Europe Reconsidered.*
96 *Ānandamath*, *BR*, i, 698.
97 *DR*, *BR*, ii, passim.
98 *BR*, ii, 643.
99 *BR*, ii, 776–7.
100 Lalita's journey to Calcutta in Tagore's novel, *Gorā.*
101 Though *Rājsinha* was written before the anuśīlan trilogy, it was practically rewritten in its second edition.
102 *KC*, *BR*, ii, 410.
103 *BR*, ii, 213.
104 Hemcandra is not a Rajput, but he is in the same cast.
105 Bankim makes references about all these three groups in his essays but wrote novels only about Rajputs. R.C. Dutt includes the Marathas. By the time Tagore writes his patriotic songs, poems and essays, all three—Rajputs, Marathas and Sikhs—are firmly established in their ironic place in the Bengali imagination, as both others and selves, as their imaginary ancestors.
106 In his remarks in the fourth edition Bankim does not show an awareness of this distinction. He gives instead moral reasons why his rewritten or reread history is true. Aurangzeb is immoral, Rājsinha is moral, thus stating quite explicitly the point about what I have called just war. *BR*, i, editor's introduction, p. 40.
107 Though Bankim does not directly invoke contractarian arguments, there is a clear influence of basic social contract ideas in his general thinking about the nature and legitimacy of governments.
108 The *Gītā*, canto 2, sloka 47.
109 Literally meaning actlessness.
110 Max Weber, *Religion of India*, passim.
111 I have dealt with this more fully in Chapter I.
112 Cf. Chapter I.
113 *BR*, i, 819.
114 Stanley Cavell, 'Must we mean what we say?' in *Must We Mean What We Say*, Cambridge University Press, Cambridge, 1976.
115 This has been wonderfully argued in Arthur Danto's *Narration and Knowledge.*
116 Danto, *Transfiguration of the Commonplace*, Harvard University Press, Cambridge (Mass.), 1981.
117 *BR*, i, 815.
118 *BR*, i, 662.
119 *BR*, i, 820.

Chapter 5

1 Compare, for instance, Vidyāsāgar's texts, *Śakuntalā* and *Sītār Banabās.*
2. Despite the fact that a number of people tried to copy his success, like Damodar Mukhopadhyay's sequels to his famous novels in *Mṛṇmayī* and *Nabābnandinī*. In a general sense, of course even authors like R.C. Dutt could be called his imitators, but they do not produce a body of artistic work which forms a totality with an internal coherence, and which could be called a distinctive form.
3 Before Bankim several distinguished writers had engaged in satirical compositions on the Bengali social world. Iswarchandra Gupta was the best known among them.

It is interesting to see Bankim's critical appreciation of Gupta's powerful, though occasionally obscene, poetry. 'Iśwar Gupter Kavitva', *BR*.

4 After the great canonical text of the *Nātyaśastra*, most others make fine distinctions between different forms of *smita* and *hasita*. The *Rasatarangini* of Bhanudatta, a text influenced by the *vyāpti* ideas of *nyāya*, makes distinctions between *vihasita, upahasita, apahasita, atihasita* etc. The also make a distinction between *svaniṣtha* or *atmastha* and *paraniṣtha* or *parastha*. Bhanudatta, *Rasatarangini*, (ed. Devdutt Kaushik), Munshiram Manoharlal, New Delhi, 1980, 128ff. But this form of *atmastha* humour is quite distinct from Bankim's self-irony.

5 Most persuasively by Reinhart Koselleck in his *Futures Past*, MIT Press, 1985, pp. 3–20.

Index